Understanding Basic Ecological Concepts

Third Edition

Audrey Tomera

revised by Joel Beller

J. WESTON WALCH PUBLISHER

Portland, Maine

User's Guide to Walch Reproducible Books

As part of our general effort to provide educational materials which are as practical and economical as possible, we have designated this publication a "reproducible book." The designation means that purchase of the book includes purchase of the right to limited reproduction of all pages on which this symbol appears:

Here is the basic Walch policy: We grant to individual purchasers of this book the right to make sufficient copies of reproducible pages for use by all students of a single teacher. This permission is limited to a single teacher, and does not apply to entire schools or school systems, so institutions purchasing the book should pass the permission on to a single teacher. Copying of the book or its parts for resale is prohibited.

Any questions regarding this policy or requests to purchase further reproduction rights should be addressed to:

Permissions Editor
J. Weston Walch, Publisher
321 Valley Street • P. O. Box 658
Portland, Maine 04104-0658

1 2 3 4 5 6 7 8 9 10
ISBN 0-8251-4265-2
Copyright © 1979, 1989, 2001
J. Weston Walch, Publisher
P. O. Box 658 • Portland, Maine 04104-0658
www.walch.com
Printed in the United States of America

CONTENTS

Introduction to Third Edition .... *vi*

1. WHAT IS ECOLOGY?
Living Things 1
Nonliving Things 4
    Activity 1: Keeping an Ecological Interaction Diary 5
Review 8

2. INDIVIDUALS, POPULATIONS, AND HOMEOSTASIS
Individuals vs. Populations 9
    Activity 2: Estimating the Density of Water Lettuce 12
    Activity 3: Measuring Classroom Density 14
    Activity 4: Investigating the Effect of Hunting on a Quail Population 15
Homeostasis—The Dynamically Balanced State 17
    Activity 5: Studying a Fruit Fly Population 22
Review 26

3. COMMUNITIES AND ECOSYSTEMS
Communities 27
Ecosystems 33
    Activity 6: Investigating the Effects of Abiotic Factors 36
Review 38

4. GOING, GOING, NOT QUITE GONE
Forest Types 39
Causes of Deforestation in the Tropics 41
Deforestation and Farming in the United States 42
Reasons for Preserving Forests 44
    Activity 7: Become a Tree Pal 45
    Activity 8: Construct a Butterfly Picnic Resort 46
Wetlands 47
Review 48

5. RELATIONSHIPS IN COMMUNITIES
Niches in a Community 49
Roles Within the Food Chain 50
    Activity 9: Constructing a Food Chain 53
    Activity 10: Constructing a Food Web 53
    Activity 11: Constructing a Food Chain in a City 54

iv   Understanding Basic Ecological Concepts

      Activity 12: Investigating Human Food Chains 55
  The Energy Pyramid 56
  Parasitism 58
  Scavenging 59
  Commensalism 60
  Mutualism 61
  Decomposers 62
  Competition 63
      Activity 13: Constructing a Food Web Showing Several
               Eater-Eaten Relationships 65
      Activity 14: Investigating Special Relationships 66
  Review 67

6. NUTRIENT CYCLES

  The Carbon Cycle 70
  The Water Cycle 72
      Activity 15: A Do-It-Yourself Water Cycle 73
  The Phosphorus Cycle 74
      Activity 16: Creating a Closed Ecosystem 76
  Review 78

7. SUCCESSION

  A Climax Community 79
  How Communities Change 80
      Activity 17: Projecting Pond Succession 84
      Activity 18: Studying Forest Succession 86
      Activity 19: Studying Succession on Your Own 97
  Review 98

8. HUMAN ECOLOGY

  Carrying Capacity 99
  Atmospheric Quality 101
  Regional Haze 102
  Smog 102
  Acid Deposition 103
      Activity 20: Where in the United States Did Acid Rain Fall in 1999? 104
      Activity 21: Studying Acid Precipitation 106
  Review 108

9. GLOBAL WARMING

  The Effects of Global Warming 111
  Causes of Global Warming 113
  The Greenhouse Effect 116

What Can We Do to Cut the Emission of Greenhouse Gases? 117
    Activity 22: Demonstrating the Greenhouse Effect 118
Review 120

10. Car Talk

Is Bigger Better? 121
Hybrid Automobiles 123
Fuel Cell Energy 124
Another Source of Energy 125
    Activity 23: Testing a Solar-Powered Model Car 126
    Activity 24: Pollution in the Business District 128
Review 131

11. Stratospheric Ozone Loss

The Ozone Cycle 134
Ozone Loss 135
    Activity 25: What Role Do They Play? 139
    Activity 26: Debating the Ozone Issue 139
Review 140

12. Water, Water, Is It Fit to Drink?

    Activity 27: The City Waterworks 143
    Activity 28: Determining Water Consumption 145
    Activity 29: The Sewage Treatment Plant 148
Review 150

13. The Solid Waste Problem: Garbage and Trash

    Activity 30: Disposing of Solid Wastes 153
    Activity 31: A Pictorial Visit to a Sanitary Landfill 156
Review 161

    Afterword: Suggestions for Further Study 163
    Answer Key 165
    Appendix I 177
    Glossary 181
    Index 191

UNDERSTANDING BASIC ECOLOGICAL CONCEPTS: INTRODUCTION TO THE THIRD EDITION

In the first edition of *Understanding Basic Ecological Concepts,* Audrey Tomera identified key ecological concerns of the time, for example, oil spills, loss of natural environments, and overpopulation. After years of study, debate, government regulation (and deregulation), and citizen action, these problems are still with us. In addition, many other environmental problems have surfaced. Acid rain, toxic waste, tropical deforestation, the greenhouse effect, and destruction of the protective ozone layer are concerns that are likely to become more important in this century.

Ecology is the branch of science that looks at organisms in their environment. In fact, the word *ecology* comes from *oikos,* the Greek word for home or place to live.

What began as the study of natural history today explores the *ecosystem,* a working unit of living and nonliving components. Ecologists look at the connections between those components, how they fit together, the way they work. This book will help students look at those connections, too, through a series of activities designed to develop critical thinking.

This new edition of *Understanding Basic Ecological Concepts* was prepared with an eye to covering the most recent developments in the field of ecology. Unchanged since Audrey Tomera wrote the first edition is the fact that "At this point in time, we are tied to life on the planet Earth."

WHAT IS ECOLOGY? 1

Ecology is a science; the people who work in this area of science are called **ecologists.** It would be easy to start with a dictionary definition of the word ecology and a short explanation of what ecologists do. But would it have much meaning to you? Probably not, so let's work at a definition of ecology in a different way.

LIVING THINGS

Below is a photograph of cattle **grazing** in a New Mexico pasture. Study the scene carefully, keeping in mind that its location—New Mexico—is quite different from, say, Michigan, Vermont, or Ontario. Many plants and animals besides the cattle live in this pasture. You would recognize some of these; others may take some thought on your part to discover. For instance, quail might nest beneath the shrubs in the background. A lizard might rest in the shade of the yucca plant. Tapeworms might be found in the intestines of the cattle. On the next page, list any animals and plants you think might live in this New Mexico pasture. Feel free to include both **organisms** that live here permanently and those that might be common visitors.

Cattle grazing in a New Mexico pasture. (Courtesy of H. R. Hungerford)

What Is Ecology?

Living Things Present in New Mexico Pasture

1. _____
2. _____
3. _____
4. _____
5. _____
6. _____
7. _____
8. _____
9. _____
10. _____
11. _____
12. _____
13. _____
14. _____
15. _____
16. _____
17. _____
18. _____
19. _____
20. _____

Ecologists are interested in studying the living things you listed, and others as well. They would study not only the grasses and shrubs but also all other plants and animals inhabiting or visiting the pasture. They would include the cattle as well as the flies, ticks, and parasitic worms that affect the cows. Ecologists would include the ants, earthworms, and beetles that burrow in the soil. The lizards and rodents that scurry among the grasses would be objects of study as well as the snakes and hawks that eat these animals. Ecologists would even study people, since the domestic cattle are evidence of people's presence in this pasture. All of these living things are also known as **biotic** factors.

Although ecologists would identify as many living things as possible, this would not be their only purpose. As you listed the animals and plants, you probably thought of them in terms of where they lived in the pasture or what they were doing there. Ecologists try to do the same. For example, burrowing insects are associated with the soil. Their burrows make air spaces in the earth, and their waste materials add nutrients to the soil. The quail and the shrubbery are related to each other also. Quail might nest beneath the shrubs and hunt for insects harmful to those shrubs.

When studying living things, ecologists might ask the following questions. Try your hand at answering them.

- In what ways are the cattle dependent on the grassland? Explain your answer. _____

- How is the grassland dependent on the cattle? _____

- How might the cattle harm the grassland? _____

- In what ways do the plants in the pasture compete with each other? _____

- What animals might compete with the cattle for food? _____

- Why might this competition occur? _____

- What is the main purpose of raising cattle? _____

- What might this environment look like in 10 years if humans and their cattle were removed? _____

Of course, grass is the main source of food for the cattle. Without it the animals would die unless humans supplied hay and silage in its place. Cattle depend almost totally on grass and other herbs. People's **dependence** on cattle, however, is only partial, since they eat other foods besides beef. Cattle can overgraze an area and destroy it as a food source. Whether they actually do this or not depends on the size of the cattle population. The waste materials from the cattle can also enrich the soil. Such relationships are the types of data the science of ecology deals with.

What Is Ecology?

NONLIVING THINGS

Ecologists help explain the natural relationships that exist between the plants and animals in our world. These relationships are called **interactions.** When organisms (biotic factors) interact, they affect one another and bring about some type of change. Cows interact with grass. One piece of evidence of this interaction relationship is a size and weight gain in the cow. Cattle manure interacts with soil, as evidenced by increased fertility of the grazing land. Ecologists' study of interactions usually includes the role of people, since people are probably the most influential change agents on earth.

There is one more item to add to your understanding of what ecology is. You are aware that the types of plants and animals found in the New Mexico pasture are very different from those found in Florida. What causes the differences? What factors might influence the kinds of living things that exist in a certain area? Give three examples below:

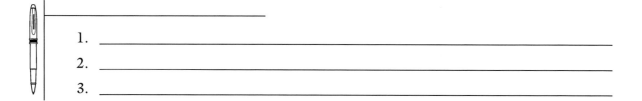

1. _____
2. _____
3. _____

If you were to study the geographic location and climate of this area of New Mexico, you would discover that it is characterized by a certain elevation, annual rainfall, temperature range, soil type, humidity, and land formations such as the mountains in the background of the photograph. Ecologists have discovered that those nonliving factors, called **abiotic** factors, greatly influence living things and interact with them. Often one abiotic factor may influence another abiotic factor, which in turn influences living things. For example, there is a decrease in temperature as elevation increases. There is an increase in **precipitation** as elevation increases. The lowered temperature and increased precipitation interact, causing a shorter growing season. These interacting abiotic factors will limit both the types and sizes of plant life able to survive at upper elevations. Soil type influences the amount of rainfall that can be absorbed. The available moisture in the soil will influence the plants that can grow there.

By putting together the three parts we have just discussed—living things, abiotic factors, and the interactions that exist between them—you should now have a good working definition of *ecology*. Check your definition with that given below.

> *Ecology:* The science that deals with the specific interactions (relationships) that exist between organisms and their living and nonliving environment.

ACTIVITY 1 | Keeping an Ecological Interaction Diary

Whenever you make a *direct* observation about something, you are on firm scientific footing. For the diary you will keep, select a tree, a shrub, a clump of grass growing in a crack in the sidewalk, a vine on a building, or a flowering plant growing in a vacant lot. Observe your plant every day for a period of one or two weeks. If you carefully (and quietly) observe your own plant specimen or grass plot for several minutes each day, you should be able to make many good observations. As you make and record your observations, try to identify what interactions are taking place. Remember that changes are evidence of interaction.

It is important to observe both living (biotic) and nonliving (abiotic) factors. For accuracy and completeness, it will be necessary to observe the subject of your investigation at various times and under different weather conditions. Time of day and weather are examples of abiotic factors. If you made your observations only at 4:00 P.M. on a sunny day when the temperature was 20°C, your investigation would not include important observations and interactions that could be seen at other times and under different weather conditions.

For example, some flowers are open only in the morning, while others are open all the time. Another example is that some plant pests, such as slugs, feed only at night and might never be seen if all your visits were during daylight hours.

To start your observations, find out if the plant is growing and at what rate. (Tips and suggestions for recording plant growth and for making other observations for this activity are shown in the box on the next page.) Listed below are some suggestions to get you started. As you make more and more visits to your plant subject, you will probably think of others.

- What animals live permanently on or in the plant?
- What animals visit from time to time?
- How do people affect the plant?
- What are the sources of the plant's moisture?
- Describe the type of soil the plant is growing in. Record the soil temperature each time you make an observation as well as the air temperature.
- What changes, if any, have occurred in your specimen's environment since your first visit?
- Briefly explain how your observations have given you a better understanding of your plant from an ecological point of view.

Tips and Suggestions

TO MEASURE PLANT GROWTH
You can use several methods. You can measure the height of the plant above the soil. Use metric measurements, because the metric system is used by scientists the world over. Another method is to put a mark with indelible ink at the base of a growing stem and another mark at the top of the stem. Then, on each visit, measure the distance between the two marks and record it in your diary.

TO SEARCH FOR ANIMALS LIVING ON THE PLANT
Bring a hand lens or magnifying glass to help you search. Many insects are quite small; you may miss seeing them if you use only your eyes.

PEOPLE AFFECTING PLANTS
As people walk near your plant subject, they may compact the soil around the plant, which would change the character of the soil. This would affect the air spaces in the soil and the amount of water absorbed.

DESCRIPTION OF THE SOIL
In order to describe the soil, pick up some of it. Is it sandy? Is there a lot of clay in the soil? Is it dark and rich with decaying material in it? Are there more pebbles than soil? There are electronic devices and other ways for measuring the moisture content of the soil and its **pH.** Many serious gardeners and schools purchase these units in order to be sure that the best conditions for plant growth are present. It may be possible to borrow one of these units for completing your plant interaction diary. Be sure to get instructions for its proper use.

MEASURING AIR AND SOIL TEMPERATURE
For these tasks, you will need a thermometer. You can use any thermometer that measures air temperature in your home or outdoors. A thermometer calibrated in the Celsius scale is the best to use, because your diary reflects scientific work and scientists record temperature in degrees Celsius. If you have only a Fahrenheit scale on your thermometer, use it and the following formula to convert Fahrenheit readings to Celsius. The formula is:

$$C = \frac{5}{9}(F - 32)$$

To take the temperature of the soil, use a stick or a thick metal bolt to make a hole in the soil. Don't use the thermometer itself as a tool to make the hole; you would probably break the thermometer. Insert the thermometer into the hole and wait two to three minutes before taking your reading. This method will be more accurate than taking a reading right away. Use a similar waiting time when taking the temperature of the air.

Sample Interaction Diary

Date: _____ **Time arrived:** _____ **Time left:** _____

Observations:

Air temperature _____   Soil temperature _____

Precipitation _____   Cloud cover _____

Wind direction _____   Wind speed _____

Other observations: _____

_____

_____

_____

Interactions: _____

_____

_____

_____

Date: _____ **Time arrived:** _____ **Time left:** _____

Observations:

Air temperature _____   Soil temperature _____

Precipitation _____   Cloud cover _____

Wind direction _____   Wind speed _____

Other observations: _____

_____

_____

_____

Interactions: _____

_____

_____

_____

Chapter 1 Review | What Is Ecology?

1. Define *ecology*. _____
   _____
   _____
   _____
   _____

2. What does the term *abiotic* mean? _____
   _____
   _____
   _____
   _____

3. List four examples of abiotic factors. _____
   _____
   _____
   _____
   _____

4. What do ecologists study? _____
   _____
   _____
   _____
   _____

5. Why is the study of ecology important to you? _____
   _____
   _____
   _____
   _____

INDIVIDUALS, POPULATIONS, AND HOMEOSTASIS | 2

INDIVIDUALS VS. POPULATIONS

If ecologists or ecology students wanted to study the organisms in a certain area such as the pasture we observed, they would have two choices. The first choice would be to study each cow, each grass plant, each specific shrub one by one. In that case, they would be studying individuals. It would be easy to do this if the subject were cows, but it would be most difficult to separate and study each individual grass plant.

The second choice would be to study all of the cows, all of the grass plants of each specific kind, all of a certain type of shrub in the area at the time of study. To investigate a **population** of organisms—a group of the same kind of individuals in a given space at a given time—would be the ecologists' usual choice. The time element is important, for a population might differ at various times of the day, during the different seasons, or from year to year. Here are two examples of populations: the number of pigeons within Chicago's city limits in 1990, and the number of corn plants per acre in Mr. Doak's field in July.

On page 10 you see a photograph of an aquatic plant called water lettuce as it floats on the surface of a Florida swamp. How many individual plants can you identify? If you wanted to study these water lettuce plants as a population, how might you go about it?

Population Density

Populations are usually measured in terms of **density.** Ecologists calculate density (D) by counting the number of individuals in the population (N) and dividing this number by the total units of space (S) the counted population occupies. The formula for calculating density thus becomes:

$$D = \frac{N}{S}$$

In studying population density on land, ecologists use the dimensions of length and width when measuring the occupied space. They therefore deal in *square units*. When studying aquatic environments, they use length, width, and depth. Space is then measured in *cubic units*.

> *Density:* The number of individuals of a particular kind of organism per unit of space at a given time.

Individuals, Populations, and Homeostasis

Water lettuce on a Florida swamp. (Courtesy of H. R. Hungerford)

Individuals, Populations, and Homeostasis

Let's look at one example of density so that we can apply the formula $D = \frac{N}{S}$. When ecologists study forest populations, they often use a space of a definite size. This space is called a **quadrat.** The size of a quadrat is usually 10 meters square (10 m by 10 m). Plots of 10 m² are marked off with twine. Within each quadrat the tree species can be plotted and counted. Suppose you are in a forest counting wild black cherry tree seedlings in a quadrat. You count 40 cherry seedlings in a particular quadrat. What is the density per square meter? $D = \frac{N}{S}$ or $D = \frac{40}{100}$. Thus, $D = 0.4$ cherry seedlings per square meter.

When studying population density, ecologists realize that density may change, often quite swiftly and dramatically. If a lumber company cut 1,000 trees in a forest, this would decrease the density. On the other hand, the germination and growth of last season's seeds would, in time, increase the density in that same forest. Both births and deaths cause changes in living populations. Ecologists call the birth rate of a population **natality** and the death rate **mortality.**

In animal populations two factors besides natality and mortality can affect density. For example, in a flock of geese on a wildlife refuge, the density could vary as more birds **migrated** in or as others left the flock for better feeding grounds elsewhere. The arrival of "new" individuals from other places to a population is termed **immigration.** The leaving of individuals from a particular population is called **emigration.**

Before you answer the following questions, here is something to consider: Start thinking of biotic and abiotic interactions that influence the population density of a flock of geese. For example, what would be the effect of a drought during the summer? That's right, it would mean fewer plants to feed the flock, resulting in a higher-than-normal death rate—particularly for the young, old, and sick members of the flock. Some members of the flock might emigrate to another area with more food. A similar effect on population density could be produced by an unusually large number of grasshoppers feeding on the plants at the same wildlife refuge.

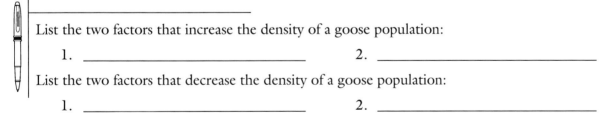

List the two factors that increase the density of a goose population:

　1. _____　　2. _____

List the two factors that decrease the density of a goose population:

　1. _____　　2. _____

The four factors that influence the density of an animal population interact to control the size of the population. As long as the variables that influence the decrease and increase in density are balanced over time, the population will remain fairly stable or balanced.

There are many factors that might influence mortality, natality, immigration, and emigration. These become more important as we study the specific interactions between living things, and between living things and the nonliving factors in their environment.

Individuals, Populations, and Homeostasis

ACTIVITY 2 | Estimating the Density of Water Lettuce

In the photograph on page 10, you viewed a population of water lettuce. Water lettuce plants will sometimes cover the entire water surface of a swamp. The plant population becomes very dense and serves as a **habitat** for many insect and spider species. Use the photograph to help you accomplish the following tasks.

1. Devise a method to estimate the number of individual water lettuce plants. Be sure to include the small plants as well as the larger ones.

    Method: _____

    _____

    Estimation: _____ water lettuce plants

2. Calculate the density of the water lettuce plants shown in the photograph. The actual space covered by the plants pictured is 2.8 square meters. Since the plants grow only on the water's surface, you need not deal with cubic units.

    Density = $\frac{N}{S}$ Density = _____ per square _____

3. Compare your density estimate with those of your classmates. What was the class average? How close was your estimate to the average? What does this tell you about the accuracy of an estimate? _____

    _____

    _____

    What method other than estimation could you use to improve your accuracy?

    _____

4. The water surface area of the entire swamp was measured by an ecologist during the same year the photograph was taken. The area is 2,500 m^2. Assume that the water lettuce covers the entire surface. On the basis of your density estimate or your more refined technique (item 3 above), what is the size of the swamp's entire water lettuce population?

    Entire swamp population = _____

5. Another ecologist was interested in a particular water spider population that lives on the water lettuce plants. A careful count of spiders found on 100 plants reveals an average of 0.82 water spiders per water lettuce plant. What is the population density of water spiders per square meter in the swamp?

   If you wished to know the population density of water spiders in the entire swamp, how would you calculate it? _____
   _____
   _____
   _____

   If the density of water lettuce plants was affected by a high winter mortality, what might be the resulting effect on the water spiders? _____
   _____
   _____
   _____

   It should now be clear to you that population-density figures benefit ecologists in understanding the types of interactions as well as the degree to which such interactions will take place in the environments they study. The study of populations, their densities, and the factors influencing the densities under study often enable ecologists to predict future environmental interactions.

Individuals, Populations, and Homeostasis

ACTIVITY 3 | Measuring Classroom Density

Human populations have density levels just like other organisms. However, people can move about, which is something some other animals and many plants cannot do. The characteristic of **locomotion** often produces interesting changes in population density.

1. Calculate the student population density of your science classroom. In obtaining the units of space occupied by the population, measure your room in meters. Remember that you will be using square meters of space.

    Density = $\frac{N}{S}$ Density = _____ per square meter

2. Repeat these calculations for each classroom you go to today. Which classroom has the highest density? The lowest? _____

_____

3. Does the population density of your science classroom remain the same each day? Graph the daily densities for one week. Plot your data on graph paper, which your teacher can supply. A sample graph is shown below.

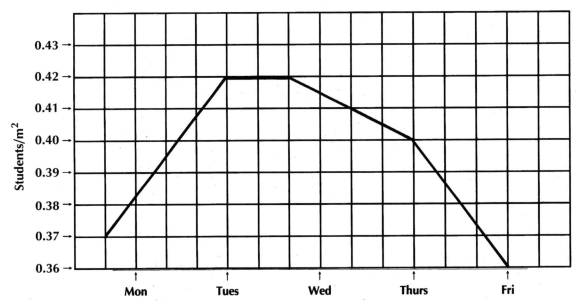

Science Classroom Population Density

4. How can you account for the daily changes in the population? _____

_____

ACTIVITY 4 | Investigating the Effect of Hunting on a Quail Population

Quail are popular game birds. Thousands of hunters shoot quail over much of the Southeast, the Midwest, and the West. There is usually a great deal of argument about the value of hunting quail, pheasant, deer, and other game animals. The following two sets of data represent the number of quail to be found on two 1,000-acre plots in one of the southeastern states in September and January during a five-year period. One game area is hunted in October and November and the other is not hunted at all. Bag limits are set by law. Graph both sets of data, using a solid line to represent the September population and a dotted line to represent the January population. Your teacher can supply you with graph paper.

1,000 Acres Being Hunted			
Year	September Population	January Population (after close of hunting season)	Quail Killed by Hunting
1984	111	41	55
1985	116	46	50
1986	110	45	46
1987	125	52	56
1988	115	40	41

1,000 Acres Not Being Hunted			
Year	September Population	January Population	Quail Killed by Hunting
1984	115	50	0
1985	120	56	0
1986	113	52	0
1987	122	60	0
1988	110	45	0

Individuals, Populations, and Homeostasis

1. Hypothesize why the quail population in the nonhunted area falls to such a low level by January each year. (Emigration is not a factor.) _____

   _____

   _____

   _____

   _____

2. Briefly give evidence (from your graphs) to support the ecologist's conclusion that regulated quail hunting is usually a harmless activity. (It may help to use the table data to determine the total population decrease in both the hunted and the non-hunted acreage.) _____

   _____

   _____

   _____

   _____

HOMEOSTASIS—THE DYNAMICALLY BALANCED STATE

It was stated earlier that populations will remain stable or balanced over time if the variables that cause the increase and decrease in density remain in balance. This idea of balance, or **stability,** is known as **homeostasis.** If you have ever watched a tightrope walker on a high wire, you have a beginning understanding of dynamic balance. The walker remains upright on the wire as he or she travels, yet may lean to the right or left in order to maintain balance throughout the journey. Homeostasis in nature operates much like the tightrope walker. Population numbers may fluctuate between highs and lows during a period of time, and yet, in the long run, these fluctuations seem to counterbalance each other and stability is maintained.

Both living (biotic) and nonliving (abiotic) factors affect changes in natality, mortality, emigration, and immigration within each population. How these factors interact is most complex. Perhaps a simplified example will begin to give you an understanding of population homeostasis.

Each fall, the Mississippi Flyway population of Canada geese migrates to winter feeding grounds in southern Illinois, northern Kentucky, and southeastern Missouri. More than 90 percent of this population winters and feeds on six wildlife refuges where a goose management program is in effect. Thousands of acres of corn, milo, and other grain crops are grown for the sole purpose of giving the geese a winter food source and feeding area. Careful counts of the population are made by wildlife biologists in order to determine whether the carrying capacity of the refuge lands is adequate for the population size. These counts are also necessary to determine a maximum goose-hunting quota that will help guarantee a sufficient number of the population to survive and breed the next spring.

Although humans are the principal factor influencing the mortality rate of the Canada goose in the winter, natural enemies such as the bald eagle and red fox also reduce the flock. The amount of food available will influence the survival rate of the geese, too. Available food is affected by the acreage set aside for food production as well as the amount of snow and ice that cover the ground. The type and amount of food needed by individual birds is determined, in part, by the winter air temperatures.

The resistance of geese to disease is influenced by the individual animal's general health. A goose's health, in turn, is directly related to the maintenance of **optimum** body weight. During a severe winter, a heavy snow cover over the food may even force some members of the population to emigrate to areas where food is more plentiful.

Since Canada geese breed in the late spring, there are no winter population additions. Emigration and immigration do not seem to influence the goose population permanently. Although some geese may emigrate to new feeding grounds during the winter, they usually migrate back to the flock as it leaves its winter residence and flies to Canadian breeding grounds.

The end result of all these interactions is a much-reduced goose population returning to its spring-summer habitat. It would seem, at this point, that the population is not in balance—

that homeostasis is not being maintained. This is not the case, however, for homeostasis is a dynamic state and cannot be viewed at any one point in time. Before we reach a conclusion, let's look at what happens to the goose population during the summer months.

A flock of Canada geese on a wildlife refuge in southern Illinois. (Courtesy of H. R. Hungerford)

During the late spring and the summer months, the Mississippi Flyway Canada goose population experiences another fluctuation. The adult geese mate and raise the young. Each breeding female raises an average of 2.5 goslings. This reproduction rate is influenced by optimal habitat. There is abundant space and food, and most of the young survive to a flying age. They will join the flock as it migrates to its winter residence. An increase in the population has thus taken place.

A large number of adult geese die during the summer and during the fall migration. Geese are affected by **predators** and by disease. Many are hunted by Eskimos. Thousands are killed during the Canadian hunting season. Others are killed by hunters in Michigan and Wisconsin as they migrate through these states on their journey to Illinois, Kentucky, and Missouri. Yet natality is so influential in this population that it offsets mortality and may even cause a slight population increase. The flock that returns to winter residence the next year is of approximately the same size as the year before.

Individuals, Populations, and Homeostasis

New populations of Canada geese have sprung up in addition to the those of the Mississippi Flyway. These are geese who now live in or near cities and towns all year. Over time, more and more humans have inhabited the Mississippi Flyway region. Homes, roads, and stores are being built on portions of the breeding grounds of the Canada geese. The geese have adapted to this situation by moving to new places and by breeding in increasing numbers. Their requirements are simple; they look for fresh water as a safe place to fly to if attacked by a land predator. Their other requirement is a food supply in the form of grass. So, many geese have taken up residence in parks, golf courses, parking lots, soccer fields, baseball outfields, shores of lakes and ponds, and even the lawns of private homes. Their presence is usually unwanted, mainly because of their distasteful digestive wastes. These wastes multiply over time and as goose populations increase.

The problem for the people affected and annoyed by the geese is how to get rid of them. Hunting sounds like an easy solution, but it isn't always practical. There are laws against hunting in city parks and within town limits. In the year 2000, officials in one town had the geese captured in nets, then transported to a wilderness area that could serve as their home. There they were released. Some golf courses have special dogs that roam the course at night chasing away the birds. Other golf courses use nets and wires strung across the grassy areas when golfers aren't playing. These devices interfere with the movement of the geese, causing them to leave the area for a better place. Many parks have laws and signs against feeding the birds. They will take a free meal of bread if it is offered. However, they can live and reproduce very nicely without a "free lunch." In extreme circumstances, homeowners could also replace their lawns with shrubs and ground-covering plants other than grass as a means of getting their unwanted guests to leave.

One of the predators that feed on Canada geese is the bald eagle. This bird, which nearly became extinct in the twentieth century, is on the **endangered species** list of the United States government. These magnificent birds were severely affected by DDT (an **insecticide** that caused the eagle's eggs to become flimsy and break, killing the chick) and hunting. By 1963, there were fewer than 1,000 bald eagles in the United States. Banning the use of DDT and prohibiting the shooting of bald eagles has begun to reverse this decline. In 2001, there were more than 10,000 bald eagles in the United States. Another factor that allowed the eagles to make a comeback was the Clean Water Act of 1972, because eagles need clean water as well as food to survive. Protecting the forests can also help the eagle, since this species needs tall trees in which to build nests. Bald eagles are not **active predators,** except as a last resort. They prefer to feed on garbage heaps outside fast-food restaurants or to steal food from weaker predators like the osprey or the cormorant rather than catching their own. What the effect of larger numbers of bald eagles will be on the Canada geese population, only time will tell.

The table on page 20 summarizes the changes that took place in the Mississippi Flyway Canada goose population for the years 1976–77 and 1987–88. In both cases, approximate figures have been given, in both numbers and interactions, to describe the situation for an

© 1979, 1989, 2001 J. Weston Walch, Publisher   19   *Understanding Basic Ecological Concepts*

Individuals, Populations, and Homeostasis

entire year. Study the figures. The table should help you understand that when dealing with populations, the following statements hold true:

1. Homeostasis must be viewed over a period of time.
2. Populations experience fluctuations while attaining stability.
3. Stability is a dynamic (changing) type of balance.
4. Homeostasis is established through an interaction of a number of living and nonliving forces.

POPULATION CHANGES IN THE MISSISSIPPI FLYWAY POPULATION OF CANADA GEESE

(All figures are approximations provided by the Technical Section of the Mississippi Flyway Council, courtesy of Donald H. Rusch.)

Activities	August 1976–August 1977			August 1987–August 1988*		
	Additions	Losses	Population Size	Additions	Losses	Population Size
Late Summer Population			450,000			843,000
Mortality by Natural Causes (Aug.–Nov.)		22,000			38,000	
Killed by Native Americans		23,000			39,000	
Hunting Harvest in Canada & Northern United States		41,000			105,000	
Maximum Number at Start of Winter Residence			364,000			661,000
Hunting Harvest (Illinois, Kentucky, Tennessee, Missouri)		22,000			64,000	
Mortality by Natural Causes (Dec.–April)		24,000			45,000	
Shotgun Crippling Loss		3,000			25,000	
Maximum Number at Start of Breeding Season			314,700			527,000
Mortality of Adults and Yearlings (May–July)		17,500			38,000	
Young Brought to Flying Stage	157,800			359,000		
Renewed Population, Late Summer			455,000			848,000

*Data for 1987–88 were provided by Professor Donald H. Rusch of the Department of Wildlife Ecology, University of Wisconsin at Madison.

© 1979, 1989, 2001 J. Weston Walch, Publisher   *Understanding Basic Ecological Concepts*

Individuals, Populations, and Homeostasis

At this point, restudy the quail population data (Activity 4) for both hunted and nonhunted areas. Are humans a necessary factor in establishing a homeostatic quail population? You should be able to cite evidence to defend your response. _____

_____

_____

So far, homeostasis has been described in terms of population changes. This is only one application of the idea of homeostasis. It also applies to the individuals that make up populations. Each organism maintains a homeostatic balance between constructive and destructive factors that influence its body. For example, in warm-blooded animals, regulatory factors help maintain a constant body temperature. If the outside temperature becomes too high, humans will perspire and dogs will pant in order to eliminate excess heat. If the outside temperature is too cold, both organisms will shiver in order to generate more body heat. These are two methods that organisms employ to maintain a steady—or homeostatic—temperature state. What other homeostatic mechanisms can you think of? Did you consider cell replacement of injured organs? Hormones such as adrenalin and estrogen? A plant's growth behavior in relation to light? A desert reptile's daytime burrowing behavior and nighttime feeding activity? Each organism must maintain a homeostatic state or risk death.

Other applications of homeostasis will become more clear to you as we study the specific interactions among living things and between living (biotic) things and the abiotic factors in their environments.

ACTIVITY 5 | Studying a Fruit Fly Population

This laboratory activity will give you an opportunity to study a population of real organisms under controlled conditions. You will watch a population grow, and you will observe the environmental changes that take place while this is happening. Subsequently, you will be able to watch the population decline, and you may determine factors that are responsible for that decline.

> **Note:** This is a long-term study in which the collection of data will take at least a month. It is important that the experiment be followed through to its conclusion.

INFORMATION ON FRUIT FLIES

The fruit fly is a well-known animal in biology because it has been used in genetics experiments for many years. It is easy to culture if you carefully follow a few instructions. The fruit fly goes through a complete life cycle in 10 to 14 days, depending mainly on temperature.

Fruit flies undergo complete **metamorphosis,** or change of form: egg, **larva** (wormlike form), **pupa** (cocoonlike stage), and finally, an adult fly. The eggs are quite small and difficult to see, although it is possible to find some with a 10× hand lens. The larvae hatch from the eggs in a day. The larvae are white and wormlike; they feed almost constantly and so grow rapidly. After approximately nine days as a wormlike form, the larva becomes a pupa. You can find the brownish pupa cases on the sides of the culture jar or on the cardboard that has been placed in the culture chamber. Adult flies emerge from the pupa cases and, in 10 hours or so, the females will mate. Egg laying will begin shortly after, and a new life cycle will begin.

> **Materials**
>
> Glass jars—2 of each size: 237 ml (½ pint), 475 ml (pint), 950 ml (quart)
>
> Ripe bananas or Instant Drosophila Medium
>
> Strips of cardboard—7 cm wide by 15 cm long
>
> Cloth squares—loosely woven and large enough to cover jars
>
> Rubber bands—to secure cloth to jar openings
>
> Yeast—dried package yeast
>
> Fruit flies—the naturally occurring wild fly or flies obtained from a biological supply house
>
> Graph paper, ruler, pencils

Setting Up Cultures

- It is recommended that two each of three different size jars be used (suggested sizes are noted on the previous page). The glass jars should be washed thoroughly in hot water, rinsed, and dried well.

- Half of a ripe banana should be placed in the bottom of each jar, together with one strip of cardboard to provide solid support for flies and a place for larvae to pupate. Sprinkle a few grains of the yeast over the banana. (This allows the fruit to ferment, a necessity for fruit fly culturing.) Cloth and rubber bands should be placed within easy reach.

- In areas where wild fruit flies are common, they can be attracted directly to the open jars. As soon as four or five flies are in the jar, cover it with the cloth and secure the cloth with a rubber band. If the banana does not attract flies, try using a piece of melon or an over-ripe peach. (You may even wish to experiment with various fruits as culture mediums.)

- An easier way to meet the nutritional needs of the fruit flies is to use a form of premade medium, which is available from biological supply houses. There are many advantages of using the "instant" medium. First, it usually comes as a dry powder. You just combine equal amounts of the powder and cold water and let the mixture stand for one minute. Sprinkle a few grains of dry yeast on the surface, and the fruit flies' food is ready to be eaten. Don't add too much yeast, because it will cause the medium to become hard. A second advantage is that the store-bought medium is mold-resistant and contains all the nutrients that fruit flies need. A third advantage is that the colored instant media serve as a good contrast for observing the pale while larvae.

- You will have more success attracting wild flies with a ripe banana than with a purchased mixture. However, if you plan to use flies from a supply company, the purchased medium will increase your chances of carrying out a successful experiment.

- If you are using commercial flies, you must transfer them carefully from the shipping container to the culture bottles. This can be done by inserting *only the mouth of the shipping bottle* into the culture bottle (between the cloth cover and the jar opening), and waiting patiently until four or five flies have entered the culture bottle. You can then remove the shipping bottle, replug it, and transfer it to the next culture bottle. Take care that flies do not escape during the transfer.

- Once you have the flies in the culture bottles, you can set the bottles aside in a warm place. However, *keep the bottles out of direct sunlight.* Tag each bottle with a date so that you can maintain an accurate record for each culture.

Collecting the Data

- A daily diary should be maintained over the life of the cultures. Be sure to record the dates of each entry. You should keep a separate record for each culture bottle. (Reasons for this will become clear as the laboratory experiment progresses.) The main reason for keeping a diary is to make daily estimates of the population of live adults in each container. However, other conditions should be recorded in the diary too. Examples of such data include the amount of waste material that has accumulated in the culture or any changes observed in the food material available to the larvae.

- The population of flies may remain static until one life cycle has been completed. In a few days, larvae should be visible in the tunnels formed as they eat their way through the food. Later, brownish pupae will appear on the sides of the bottle and on the cardboard strip. New flies should begin to emerge from the pupae within two weeks. Count the flies in each container every day and record the number in your diary. As more and more flies appear, you will only be able to make *estimates* of the fly population in each culture bottle. These estimates should be as accurate as possible.

- As the days pass, the fly population should become extremely large. Soon you will notice changes within each culture bottle. There will be dead flies. These may be replaced by new flies for a time, but sooner or later the fruit fly population in each container will begin to decline. Watch this and subsequent developments very carefully. When the population in a container reaches approximately 25 percent of the size it was at its height, stop the experiment. Record the date and the population size in the appropriate place in your diary. Then take the fly container outdoors and release the remaining flies, or transfer them to a fresh culture dish with ample food and no accumulation of wastes if you wish to repeat this experiment or perform a different experiment. *Don't release the flies inside your house or school!*

- Once all the containers have been emptied of flies and the experiment is complete, enter the data on graph paper, which your teacher can supply. Prepare a separate line graph for each container. Use the horizontal axis to record time (days or weeks) and the vertical axis for population numbers. By graphing the data for each size container, you will have pictures of each population's growth and decrease. You can then make inferences regarding the factors that determine the population increases and declines in each closed fruit fly environment.

- Since you stopped the experiment before the fly populations dropped to zero, the first thing to do with your graphs is to extend each curve to the right, following its normal slope until it reaches the zero point. Indicate the day on the graph and in your diary when you predict that all the flies would have died if you had continued to experiment.

Questions to Consider

1. Describe the pattern of each growth curve. Did it rise sharply or gradually? Did it come down the same way it went up? Was there a period of time when the curve was level? When did this occur? Compare all the graphs: What effect did the size of the culture container have on the shape of the growth curve? _____

    _____

    _____

2. What abiotic changes did you observe within the culture containers as the populations fluctuated? _____

    _____

    _____

3. Did a homeostatic state exist in any of the fly cultures? Explain your answer.

    _____

    _____

    _____

4. Would the population growth curve for the closed fruit fly culture be the same as a fruit fly population in the wild—that is, outside the culture bottle? Justify your answer.

    _____

    _____

    _____

5. Populations are limited by numerous environmental factors. What abiotic factors seemed to limit the populations in the culture containers? (It is important to remember that there was no immigration or emigration!) Explain the population changes in terms of natality and mortality. _____

    _____

    _____

6. How do your findings about fruit flies relate to population growth and/or decline in other populations, such as forest trees, deer, algae, or human beings? For example, you noticed a decline as the culture medium became contaminated. How might various types of **pollution** affect other kinds of populations? _____

    _____

    _____

Individuals, Populations, and Homeostasis

Chapter 2 Review | Individuals, Populations, and Homeostasis

1. Name four factors that affect the density of populations. _____

2. If other factors remain constant, what effect will emigration have upon population density?

3. Why is it wiser to study populations rather than individuals in an ecosystem?

4. What does the term *homeostasis* mean? _____

5. What effect will the increase in the bald eagle populations probably have on the Canada geese population? _____

COMMUNITIES AND ECOSYSTEMS 3

COMMUNITIES

The photograph of the New Mexico cattle pasture on page 1 represents one example of a community. Five other photographs of communities appear below and on following pages. Observe each photograph carefully. You will find it rather easy to observe differences. Although differences are important, it is the similarities that are most important here. The similarities may be harder to note. Still, try to determine what the photographs have in common.

Arizona desert community. (Courtesy of H. R. Hungerford)

Communities and Ecosystems

Florida pine flatwoods. (Courtesy of H. R. Hungerford)

Emergent pond community. (Courtesy of H. R. Hungerford)

Bog community. (Courtesy of H. R. Hungerford)

Common milkweed flower head. (Courtesy of H. R. Hungerford)

Communities and Ecosystems

Note below the similarities you believe exist among these five photographs.

_____
_____
_____
_____
_____

All five photographs represent what could be called **natural biological communities.** From your observations, how would you define a natural biological community? Write your best definition below.

_____
_____
_____

A natural biological community is really any relationship in nature that involves plants and animals living and interacting together. The community can be a forest, a **prairie,** a pond, a bog, a desert, or an ocean tidal zone. What other natural biological communities can you think of?

When you were asked to define a natural community, you probably had a pretty good idea until you saw the milkweed flower head. This photograph might have caused you to stop and wonder if your ideas were right. You may still think that this is a bad example of a community.

Natural communities do not have to cover large areas of the earth's surface. Nor do they have to be long-lived (although many are). A milkweed flower head may last less than a week, but during that time it becomes a teeming community structure. The flower head is visited regularly by a variety of bees, and several kinds of beetles live on it. Tiny insects called thrips may have populations of up to two hundred on a single flower head. Ants may be found by the dozen. Spiders often take up living areas in milkweed flowers to prey on bees and other insects. The monarch butterfly may be the best known milkweed insect; it lays its eggs right on the milkweed plant.

A **community** to ecologists is more of an idea than anything else. We call that idea a concept. A concept is merely the mental image we have of something or some relationship we are familiar with.

The natural biological community is a working concept for ecologists. It is sometimes referred to as the community concept. The community concept gives us a way of looking at ecology in an orderly, logical manner. It permits us to study ecological conditions and interactions in an organized fashion. Thus, we study the forest, or the pond, or the milkweed flower head as a community of organisms that interact with one another.

The photo essay that follows on the next pages should help you develop a concept of some of the interactions that take place in a prairie community.

A Grassland Prairie Community

This is a small section of prairie in Oklahoma. The key plant in this community is grass. The grass, in turn, depends to a great extent on the soil in which it grows.

Grazing animals are dominant members of the prairie community. The buffalo, once a dominant grassland consumer, still lives here protected. (Courtesy of H. R. Hungerford)

Communities and Ecosystems

The dung or solid waste of large animals is important in recycling nutrients back to the soil of the prairie. It also becomes the home of many organisms, including insects and bacteria.

The crab spider is a predator on the prairie. Here the spider has killed a honeybee that came to the milkweed flower seeking nectar. What other predators probably live on the prairie? (Courtesy of H. R. Hungerford)

ECOSYSTEMS

When we look at natural communities, we are observing populations of organisms interacting with one another. Because all parts of the community are alive or were living at one time, such a community is referred to as a *biotic system,* full of living—or biological—factors.

Some ecologists feel that we cannot intelligently study communities if we ignore their nonliving factors. This argument is very sound, because many abiotic factors greatly influence living communities. This is why the term **ecosystem** came into existence. A more complex and complete concept than the community, the ecosystem encompasses both the living and the nonliving factors in a particular **environment.**

Think for a minute of the many abiotic factors in an ecosystem, besides temperature and rainfall, that might influence communities of living things. List your ideas below.

Abiotic Factors

1. _____
2. _____
3. _____
4. _____
5. _____

6. _____
7. _____
8. _____
9. _____
10. _____

Some of the factors you might have listed are: light, minerals, wind, humidity, elevation, gravity, the medium upon which the organisms exist (water, mud, sand, rock), predominant land forms, wave action, and tides. Some of these physical factors are more important than others in influencing the nature of the community. And organisms differ in the extent to which they are controlled or limited by various physical factors.

Let's look at two examples of how abiotic factors can influence various communities. Take a close look at the following photograph. It shows a desert countryside in southwestern Arizona. Something about the picture should seem inconsistent or unexpected. What is it?

Communities and Ecosystems

North- and south-facing slopes in Arizona. (Courtesy of H. R. Hungerford)

A close look at the photograph reveals a very different living community on one slope than on the other. The south-facing slope (the one to the right) receives a great deal more sunlight than the other. Therefore, the evaporation rate of water is much higher there. Only plants that can exist in the driest conditions are found here, growing in the direct rays of the sun. Because the slope facing north retains more water, it can support shrubs and herbs that cannot survive on the other slope. Which slope do you suppose contains the toughest, most highly adapted plants? Which slope serves as soil for the richest biotic community? Why? Both slopes can be considered parts of the desert ecosystem, since we are concerned with abiotic factors also.

Every community is influenced by a particular set of abiotic factors. These abiotic factors affect the community members. However, the living members may also influence the abiotic factors. For example, the amount of water lost through the leaves of plants may add to the moisture content of the air. The leaves of forest trees reduce the amount of sunlight that penetrates the lower regions of the forest. The air temperature is, therefore, much lower than in nonshaded areas. What additional examples can you give?

In the beginning of this chapter, you viewed photographs of several different communities—a pond, a desert, pine flatwoods, a bog, and a milkweed flower head. Study each of these communities again. This time consider each as an ecosystem. Try to identify the abiotic factors that exist in each ecosystem, and determine how these factors influence the living members.

In order for a community or an ecosystem to exist, a **dynamic** balance must be maintained among all living and nonliving factors. Yes, the concept of **homeostasis** applies not only to individuals and populations but also to the total ecosystem. Ecologists study an ecosystem not only to determine what living and nonliving factors make up the system, but also to understand how all the factors interact to maintain the delicate balance that allows the ecosystem to exist at all. This is the basic business of ecology.

ACTIVITY 6 | Investigating the Effects of Abiotic Factors

There are many ways in which you can study the effects of abiotic factors firsthand. The following three suggestions may work well for you, but if you should develop other ways of observing the effects of abiotic factors, it is recommended that you use those methods.

Observing the Effect of Light Energy on Aquarium Water

Obtain two or four 4-liter or 1-gallon glass jars. Wash and rinse them thoroughly. Pour 2 liters (approximately half a gallon) of pond water or water from a well-established aquarium into each. Place a piece of glass or plastic over the opening of each jar. Put half of the jars in a place where they will receive a great deal of light energy, such as a south window or under a gooseneck lamp. Place the other half in a shaded area, away from intense light energy. Observe all jars over a period of several days. What differences do you observe? Using a thermometer, how do you compare the temperature in the sunlit jars with the temperature in the shaded jars? Which of your observations might be due to temperature differences? Which observations might be due to the light conditions?

Observing the Effect of Automotive Exhaust on Roadside Plants

Locate a road that is heavily traveled by many cars and trucks and has a border of soil with the usual hardy wild plants—such as dandelions, crabgrass, and perhaps some wildflowers. Survey the number of plants growing closest to the concrete or blacktop. Count the number of plants in a strip measuring 0.25 meter by 2 meters. Observe their height and their general state of health. What species of plants are growing at the edge of the road? Next, make observations in a strip that is identically sized but about 1.5 meters away from the road. What differences do you note? What inference can you make about the effect of car and truck exhaust on green plants?

Observing the Effect of Shade on Biotic Conditions

If possible, find a building that has a lawn growing up to both the south and the north walls. (A school might prove an ideal place for this investigation.) Observe and record all biotic (living) and abiotic (nonliving) conditions at the point where the lawn meets the wall of the building.

Besides measuring temperature and light intensity, you might attempt to get some rough moisture comparisons through simple touch techniques. What kinds of organisms are present? If the same animals are found on both sides of the building, are they found in the same numbers?

The boxed Observation Tips on page 37 will help you in your investigation. They will also help you identify some common soil organisms.

What organisms are in the soil itself on both sides of the building? If you can get permission, dig out a 25-centimeter (10-inch) square piece of sod to a depth of about 10 centimeters (4 inches). Do this on both sides of the building. Collect the sod and soil in plastic bags that can be sealed to prevent the loss of moisture. On an old sheet or white paper spread over newspaper, look through your soil samples to determine the kinds and number of organisms that are present. What differences in populations can you observe? What abiotic factors influence these two communities?

OBSERVATION TIPS

Observing the Effect of Light Energy on Aquarium Water

Too much light energy will be converted into heat, and many of the one-celled animals will perish. Algae, on the other hand, may thrive under the bright light conditions.

Observing the Effect of Automotive Exhaust on Roadside Plants

Species of plants will vary from one area to another. Plants growing closest to the automotive traffic should be smaller in size, fewer in number, and fewer in total number of species. Most observers will infer that automotive exhaust is not good for the health and growth of plants.

Observing the Effect of Shade on Biotic Conditions

Earthworms and beetles will be easy to find. Mites will also be present in large numbers. They are related to spiders and can be identified by their four pairs of walking legs and large, unsegmented abdomens. They look like tiny spiders. Springtails will be found in large numbers, but they will not be quite as numerous as the mites. Springtails have a forked structure called a *furcula* in the rear of the "belly." They use the furcula to spring into the air. Springtails are insects because they have only three pairs of walking legs. Look for pseudospiders, which can be recognized by their clawlike pincers and segmented abdomens. They, too, are tiny arachnids and have four pairs of legs for walking.

This list is by no means complete. If the identification of all the organisms you find is important to you, it would be wise for you to consult a book devoted to identifying insects and spiders.

Chapter 3 Review | Communities and Ecosystems

1. How does an ecosystem differ from a community? _____

2. How is a desert community different from a bog community? _____

3. What kind of predators might you find in a desert community? _____

4. Why are the spines on cacti an important adaptation? _____

5. What are the abiotic factors most necessary for a prairie to exist? _____

GOING, GOING, NOT QUITE GONE | 4

FOREST TYPES

Not only are many plants and animals considered endangered species, but we also have to consider entire ecosystems as endangered. Let's start with forests. The ones you are most likely to read or hear about are **tropical rain forests.** Actually, at least half of the forests on the earth are located in the tropics. When you read the term *tropical rain forest,* you probably think of thick jungles with very tall trees and thick vines, as well as shrubs and small trees growing under the taller ones. In fact, the picture on the right is what a rain forest actually looks like. It is called a rain forest because it may receive 1,000 centimeters of rain each year.

Tropical rain forest. (Corel CD)

How many centimeters of rain fell last year where you live? _____

The weather in tropical rain forests is hot, very humid (70% or more), and wet. The temperature varies between 20 °C and 35 °C. The largest rain forests are in Brazil and the Congo region of Africa.

A **mangrove forest** is a special type of rain forest. It is composed mainly of mangrove trees, which only grow in swampy places between the ocean and the shore. There are also **dry tropical forests.** These forests are found in India and Australia. Their name comes from the fact that they get only 50 cm of rain a year; their trees are therefore adapted to drought conditions. The dry tropical forest trees are smaller than those found in the rain forests.

Between the dry tropical forest and the grassy plains, the **savanna** can be found. In a savanna, the trees are widely scattered. If a dry tropical forest area is subject to heavy grazing by animals or to many forest fires, the dry forest changes to a savanna. On the other hand, a savanna can become a dry forest if it is exposed to minimal grazing by animals and few fires.

Going, Going, Not Quite Gone

Until the last quarter of the twentieth century, the world's tropical forests covered 15.3 billion acres. Between 1985 and 1990, however, 210 million acres were destroyed according to one group of ecologists. Scientists of the World Wildlife Fund, using data from several sources, estimated that between 50 and 60 percent of the Amazon jungle in Brazil was **deforested** by the end of the year 2000. Using the lower estimate (50 percent), this means that an astounding 7.65 billion acres have been deforested.

Just as with the issue of global warming, there are some ecologists and politicians who believe that the world's tropical forests are not in any danger. They state, according to other figures, that the loss is only 7.5 million acres per year. They claim that their studies and **satellite images** show that less than 2 percent of the Amazon rain forest in Brazil is deforested yearly. There is quite a difference between 7.65 billion acres per year and 225 million acres—and quite a difference between 50 percent and less than 2 percent. Whose numbers are correct? Or are they both wrong? And, even assuming that the lower figure of 7.5 million acres per year is correct, is this an acceptable amount of forest to lose?

CAUSES OF DEFORESTATION IN THE TROPICS

Tropical forests are cleared by the local inhabitants for **agriculture** using a technique called "slash-and-burn agriculture."

What does this term mean? _____
_____

After the forest area has been cleared and the planting of crops has begun, the **topsoil** is exposed to the sun and becomes hard. The heavy rains of the Tropics will wash the minerals out of the soil. After a few years, no crops will grow in the cleared area because the soil has lost its nutrients. So, that farm is abandoned and a new forest area is cleared. It may take as long as 100 years for the abandoned farm to become a tropical rain forest again. Larger farms in the tropics sustain sugar, cocoa, and other crops. These cleared areas are maintained year after year using fertilizers that are added to the soil.

During the late 1990s, research scientists in Costa Rica discovered that the rising temperature due to global warming has slowed down the growth of tropical trees. As the growth of these trees slows down, so does the process of **photosynthesis.** The trees therefore take in less carbon dioxide. The result is a deadly cycle: As the temperature rises, tropical forest growth and photosynthesis slow down. More carbon dioxide builds up in the **atmosphere.** This, in turn, causes more global warming, which will again lower the rate of photosynthesis. The cycle will go on and on and on if left unchecked.

People living near tropical forests need wood for cooking fires and other purposes. The result is more deforestation. Dry forests suffer the most from this problem and are often changed into grasslands.

All over the world, people want furniture made from tropical hardwoods such as mahogany and ebony. The best and largest trees are cut down with chain saws. As the trees fall, what will happen to the younger trees nearby? _____
_____

Add to this loss the fact that other trees have to be bulldozed to make roads to haul the logs out. The logs are shipped to countries all over the world.

Eight thousand tree species are close to **extinction.** Most of them grow in the Tropics. They are threatened by human activities, uncontrolled forest fires, and **competition** from foreign species that have been transplanted from other ecosystems. A tree on the endangered species list that might remind you of a Christmas story is the frankincense tree. Another endangered tree that is native to the United States is the giant sequoia.

Going, Going, Not Quite Gone

DEFORESTATION AND FARMING IN THE UNITED STATES

Let's leave the Tropics and turn our attention to forests in the northeastern United States. When the Pilgrims landed on the east coast, about 45 percent of the land was forest. Today forests cover about 30 percent of the land.

None of the trees that were here when the Pilgrims arrived are still alive. Why not?

_____
_____
_____
_____
_____

Even before the outbreak of the Civil War (1860), people were worried that too much of the northeastern woodlands were being deforested for farming or logging. Destruction of these forests continued until 1920 or so. Then the need to destroy more northeastern woodland forests for farms ended.

Examine the graph that follows. Using data from the graph, answer the questions below.

Giant Sequoia.

1. How many acres of forest land were there in 1860? _____

2. How many acres of farmland were there in 1860? _____

3. Compare the amount of forest acreage in 1930 to the amount of land used to grow crops.

_____
_____

4. Why are both lines in the graph almost flat from 1930 to the year 2000? List as many reasons as possible before reading the next paragraph.

_____
_____
_____

Going, Going, Not Quite Gone

AMERICAN FORESTS AND FARMS, 1860–2000

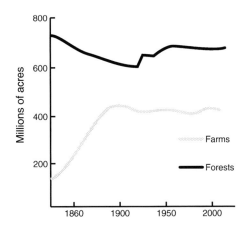

As civilization moved west in the nineteenth and early twentieth centuries, the grasslands and prairies in our nation's Midwest became the new farms. After about 1920, however, this increase in farmland began leveling off. One reason for the flat graph lines between 1930 and 2000 was that, during these decades, horses and mules on farms were replaced by machines. More crops could thus be grown—and harvested faster. Better methods of fertilization, **irrigation,** and genetically improved plants meant that the existing farmland could produce enough food.

5. At about the same time, wood for fires to heat homes and businesses—as well as fires for cooking—was replaced. Can you list three of these substitutes for wood?

_____

_____

_____

Many forests in the Northeast actually began to recover and eventually grew back to their original size. Some actually increased in size and covered more acres.

6. How do you think the development of plastics has helped preserve our forests?

_____

_____

7. What role might computers and e-mail play in forest preservation? _____

_____

_____

Going, Going, Not Quite Gone

REASONS FOR PRESERVING FORESTS

In addition to their beauty, trees provide many other benefits to their communities. They strongly influence the climate of the ecosystem they live in by shading the forest floor and helping to moderate temperatures. Trees, the largest of all plants, also help create an ecosystem in which mammals, reptiles, insects, and birds can live at the same time. Large mammals and birds are the first to be wiped out when a forest is destroyed. Next to go are the small mammals and insects.

By means of photosynthesis, trees provide food for animals. They also return oxygen to the atmosphere so that living organisms can carry on respiration. (This is explained in the section on the carbon cycle; see page 70.) Trees and other plants also filter contaminants from their ecosystems. For example, sunflowers remove radioactive substances from the soil. The Indian mustard plant has reduced the amount of poisonous lead in the soil of New Jersey. Trees filter out air pollutants, absorb noise, and cool you in the summer by providing shade.

Tree roots hold the soil in place during floods and remove water from the soil in order to live and grow. (See the section on the water cycle on page 72.) Tree leaves and bark provide us with medicines. For example, the Pacific yew was considered a weed (any unwanted plant) until it was discovered that its bark contains a compound that is very helpful in treating several types of cancer.

Most of the endangered species in the world live in the Tropics. Brazil, with its huge tropical forests, leads the world in terms of being home to a staggering number of endangered species: 55,000 different plant species; 54,000 mammal species; 16,000 bird species; 525 species of amphibians; and 525 reptile species.

Consider what is happening to the monarch butterfly. Each year millions of these butterflies leave the United States and winter in one particular part of the forest covering the mountains of central Mexico. Airplane photographs taken since 1970 show that loggers have reduced that special forest area by 50 percent. This has happened despite the fact that the area was set aside by the Mexican government as a **preserve** for the monarchs. If the logging continues, that forest will be completely gone in another thirty years, and so will the monarch butterflies.

A positive step toward preserving the world's forests was taken in the summer of 2000 by the government of Gabon in Africa. Gabon agreed to keep chain saws and bulldozers out of a 1,900 square mile **reserve.** This was good news for the trees as well as the elephants and large apes inhabiting that tropical rain forest.

Worldwide, similar steps must be taken. Leaders must decide together how much forest can be destroyed yearly for commercial reasons. Strong measures must also be taken to promote forest regrowth.

Laws for protecting the forests must be enforced by the nations of the world. Do you think these measures would work? Explain your answer. _____

_____

_____

ACTIVITY 7 | Become a Tree Pal

1. Select a tree in your area that appeals to you. It could be in a park, your front lawn, or somewhere else in your neighborhood. You might want to pick a tree that needs a friend. It may have been hit by a car, attacked by insects, or abused by vandals. In many cities, trees are dying faster than newly planted ones can grow.

2. Keep a written tree record.

   - Use a tree guide or other library source to identify the species of your tree.
   - Make a hand-drawn sketch of the tree or take its picture with a camera.
   - Collect a small sample of its leaves, flowers, fruits, and bark. **(Only collect bark that has fallen off the tree. Do not peel the bark off the tree.)**
   - Observe your tree weekly. Record your observations. What birds visit the tree? Are insects feeding on its leaves, branches, or trunk? What kind(s)? What changes have occurred to its branches since your last visit? Look for signs of tree abuse (carvings into its bark, motor oil poured at its base, etc.). Does the tree seem to receive sufficient water? If not, is there a way you could water it? What changes have taken place in its leaves/needles?
   - Discuss with a local botanist how you can help your tree if it has been attacked by insects or abused.

Going, Going, Not Quite Gone

ACTIVITY 8 | Construct a Butterfly Picnic Resort

Butterflies feed on the nectar in flowers. Like us, butterflies prefer certain foods. Therefore, some flowers are especially attractive to butterflies because of their nectar.

There are more than 700 species of butterflies in North America. Building this refuge may provide you with the opportunity to see monarch butterflies before they become **extinct**.

1. Visit a local garden supply store. Ask which of the following plants will grow where you live: phlox, lilac, aster, and mint. Get planting directions, including the best type of soil to use. Find out how deep to plant the seeds and when to plant them.

2. In a small section of your garden or in several large containers, plant some varieties of phlox, lilac, aster, and/or mint. Note: If you live in an apartment, use a large window box or a couple of small ones. (Plant the seeds of species that won't grow into tall plants. Asters and mint would be good choices.)

3. Tend your garden carefully as the seeds **germinate** and grow into healthy plants. You will be rewarded, perhaps, by a visiting monarch or two. Record your observations of any visits these beautiful creatures make to your butterfly-friendly garden.

WETLANDS

Forests aren't the only ecosystems that are threatened. The world's **wetlands** are also in trouble. Wetlands along the coast contain water that is **brackish** (containing salt). In the middle of the United States, the wetlands usually contain fresh water. Some people think that wetlands are just places where mosquitoes breed. Actually, wetlands serve many useful purposes. They provide a home for birds, fish, and animals including frogs and snakes. They protect against floods. Inland wetlands are even sources of drinking water.

Humans present the biggest threat to wetlands. As the population increases, builders want to fill in wetlands to build homes and stores. People need water in order to live, so the water in the wetlands is removed for use in nearby towns and cities. This is what happened in the Florida Everglades, which are quite large—more than 10,000 square kilometers. Slow-moving fresh water has always been responsible for

Egret standing in a marsh.

keeping the Everglades a marsh. This ecosystem is now in danger, because much of its fresh water has been piped to nearby housing developments and farms. Moreover, in order to change parts of this marsh to dry land for building purposes, canals have been built. Billions of gallons of fresh water have been drained from the Everglades through these canals and deposited in the ocean and bays. It was recently estimated that 1.7 billion gallons of fresh water were being lost to the plants and animals of the Everglades in this way daily.

Starting in 1999, the federal government and Florida state government began working on a six-year project to restore as much of the Everglades as possible. Plans for the project involved removing some of the canals and changing other procedures to help begin restoring this valuable ecosystem.

CHAPTER 4 REVIEW | Going, Going, Not Quite Gone

1. Why are tropical forests being destroyed? List as many reasons as you can.

   _____

   _____

   _____

   _____

2. Why are wetlands being destroyed? List as many reasons as you can.

   _____

   _____

   _____

   _____

3. Why should tropical forests be preserved? List as many reasons as you can.

   _____

   _____

   _____

   _____

4. Why should wetlands be preserved? List as many reasons as you can.

   _____

   _____

   _____

   _____

5. What can you do yourself to help preserve either or both of these ecosystems?

   _____

   _____

   _____

   _____

RELATIONSHIPS IN COMMUNITIES 5

NICHES IN A COMMUNITY

The natural Oklahoma prairie community you observed in Chapter 3 is a very complex one. You were probably able to see only about a half-dozen different prairie organisms in the photographs. In fact, there are dozens—even hundreds—more. The same holds true for any large community—ponds, forests, swamps, rivers, deserts, lakes, and so forth.

We noted earlier that a community involves plants and animals living together and interacting with one another. If we observe that hundreds of different species live together in a given area, we might suspect that many kinds of relationships exist there. Each type of community member has a certain role or job to fulfill in the community. Ecologists call this role a **niche.**

The figure on page 51 illustrates some of the niches that exist in a redwood forest community that one might find in the western United States. The entire community seems to be dominated by one major activity. What activity is it? _____

If you believe that food-getting is the major activity, you are right. All plants and animals must have **energy** in order to carry on their life tasks and maintain homeostasis. Energy is available only through those chemical compounds we call "food." Of course, the food sources vary. Green plants can make their own food. They represent the food **producer** niche. Some animals live directly off the green plants; they fulfill the **herbivore** niche. Others eat other animals—the **carnivore** niche. **Parasites** set up housekeeping on—or in—a plant or animal and live off the food obtained by that organism. **Scavengers** use dead plants and/or animals, as do the **decomposers,** which help break down the wastes and dead bodies of all other organisms.

The original source of all energy going into food is the sun. This is because plants that have **chlorophyll** are able to combine water and carbon dioxide in the presence of light energy and produce sugar. This sugar can be converted to energy as the plant needs it. Of course, some of the sugar is converted into other complex chemicals that permit other life processes.

A community, then, is really a group of plants and animals living together and obtaining food energy from somewhere in the community. Each community species assumes a particular food niche. Of course, there are some exceptions to this. A bird, for example, may nest in one community and obtain its food in another. But the exceptions are few in number compared with the vast numbers of plants and animals that inhabit a given community.

Relationships in Communities

ROLES WITHIN THE FOOD CHAIN

It is safe to say that food energy passes through a community in various ways. Each separate way is called a **food chain.**

One prairie food chain can be seen in the drawing at the top of page 52. The grass manufactures food. The grass is called a food producer. The grass is eaten by a prairie dog. Because the prairie dog lives directly off the grass, it is termed a first-order consumer. A weasel may kill and eat the prairie dog. The weasel is, therefore, a predator and would be termed a second-order consumer. The second-order consumer is *twice* removed from the green grass. The weasel, in turn, may be eaten by a large hawk or eagle. The bird that kills and eats the weasel would therefore be a third-order consumer, *three* times removed from the green grass. Of course, the hawk would give off waste materials and eventually die itself. Wastes and dead organisms are then acted on by decomposers. Decomposers help make the body tissues or wastes available for reuse as nutrients by the green plants while they secure food energy for themselves. Examples of decomposers are molds, bacteria, mushrooms, and other **fungi.**

It is evident that grass isn't eaten only by prairie dogs. Which other animals might eat grass and be first-order consumers? Of course, the prairie dog might also be killed and eaten by animals other than weasels. The whole situation can become very complex. When we attempt to identify the complex food relationships among many animals and plants within a community, we find it useful to create something called a **food web.** Food webs can be considered as an interrelated network of individual food chains. They give ecologists a much more realistic picture of the energy relationships in a community.

Study the incomplete prairie food web shown on page 52. What organisms other than those shown might actually exist in this food web? What would probably happen to this food web if the grass on the prairie were suddenly to disappear? What would happen if everything remained the same except that the decomposers disappeared?

_____
_____
_____
_____
_____
_____
_____
_____
_____
_____

Understanding Basic Ecological Concepts

Relationships in Communities

Niches in a Redwood Forest Community

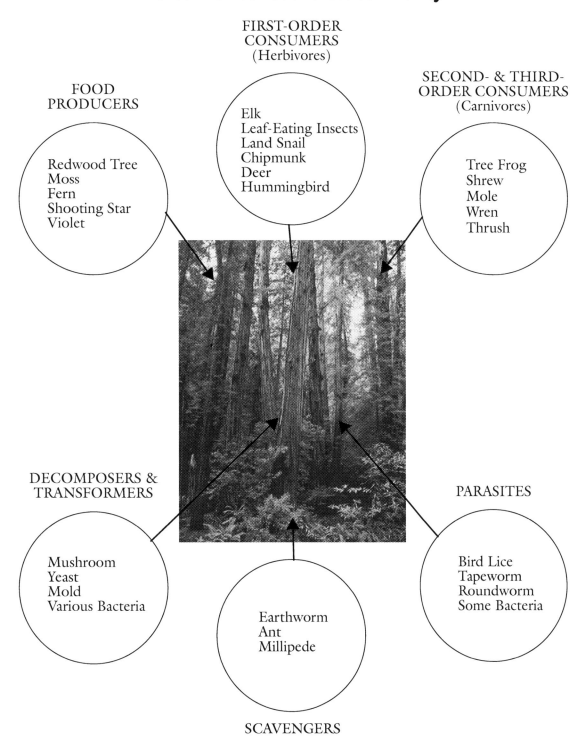

FOOD PRODUCERS
- Redwood Tree
- Moss
- Fern
- Shooting Star
- Violet

FIRST-ORDER CONSUMERS (Herbivores)
- Elk
- Leaf-Eating Insects
- Land Snail
- Chipmunk
- Deer
- Hummingbird

SECOND- & THIRD-ORDER CONSUMERS (Carnivores)
- Tree Frog
- Shrew
- Mole
- Wren
- Thrush

DECOMPOSERS & TRANSFORMERS
- Mushroom
- Yeast
- Mold
- Various Bacteria

SCAVENGERS
- Earthworm
- Ant
- Millipede

PARASITES
- Bird Lice
- Tapeworm
- Roundworm
- Some Bacteria

Understanding Basic Ecological Concepts

Relationships in Communities

PRAIRIE FOOD CHAIN

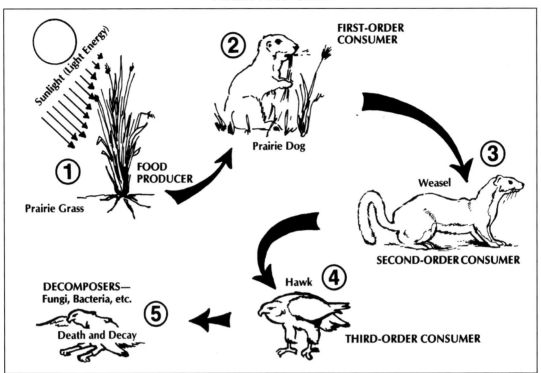

INCOMPLETE PRAIRIE FOOD WEB

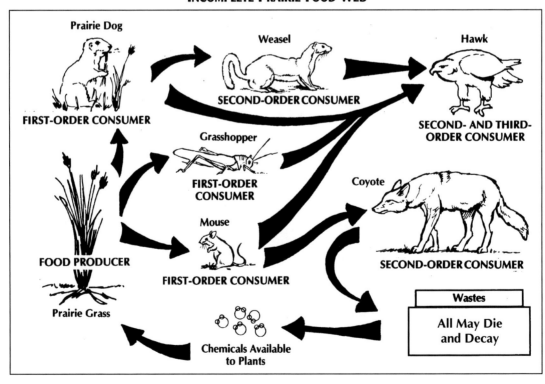

Relationships in Communities

ACTIVITY 9 | Constructing a Food Chain

Using the illustration of a redwood forest community (page 51), draw and label a food chain below. Include in this food chain a producer and at least three consumers. Label each step in the food chain, and identify each organism according to its place (niche) in the chain. Draw arrows between each organism. The arrow should point toward the organism that receives the energy.

ACTIVITY 10 | Constructing a Food Web

Determine how the food chain you have just completed above could be changed into a food web. Using a colored pencil, introduce two new organisms into the food chain so that it becomes a food web. Do all new work with the colored pencil so you can easily see the change from a food chain to a food web. Label each new organism.

Relationships in Communities

ACTIVITY 11 | Constructing a Food Chain in a City

When people build a city, they change the environment to a great extent. What kind of "natural food chains" exist in the city? Where do the sparrows, bees, mosquitoes, moths, flies, spiders, cockroaches, rats, mice, garter snakes, chimney swifts, robins, and martins get their food?

Construct a food chain that might exist in a city—perhaps in or near a vacant lot, a park, or even a crack in the concrete where plants are growing. If possible, use one producer and at least three consumers. How does the food chain differ from the one you drew before?

How do humans affect the ecology of a city area?

Relationships in Communities

ACTIVITY 12 | Investigating Human Food Chains

At the left below, you will find a drawing of a common food chain, showing a person as a first-order consumer. Many millions of people fit into this food niche. In the space to the right, draw a similar food chain showing a person in a second-order consumer niche.

Because people use short and seemingly efficient food chains, they are seldom third-order consumers. What is an example of a person as a third-order consumer?

_____
_____
_____

THE ENERGY PYRAMID

Food chains and webs explain how energy moves through a community. They do not, however, explain how much energy passes from one step in a food chain to another. Unfortunately, organisms are not very energy-efficient. Each organism uses a great deal of the energy it takes in just to stay alive, grow, and maintain its homeostatic balance. This is why food is necessary in the first place.

While it is undergoing its life functions, each organism loses energy to the environment in the form of heat. Because of energy loss at each step of food-energy transfer (the food chain), the amount of energy available to the next level in the chain is decreased.

Ecologists use an energy **(food) pyramid** concept to help show the quantity of energy that passes through organisms in the environment. Study the drawing of the energy pyramid on page 57. As represented in this diagram, the number of single-celled **algae** (producers) runs into the millions; the crustaceans (first-order consumers) that eat the algae number in the thousands. The number of sunfish (second-order consumers) that feed on crustaceans is much less. Finally, the number of great blue herons (third-order consumers) that the sunfish can support may be only one. Because of energy loss, fewer numbers of organisms can be supported at each succeeding food level.

The energy pyramid concept should help you understand why there are so few whales in the earth's oceans when compared with the billions of plankton organisms on which whales feed. Why does it take so many acres of grass to support one cow? Or, why are there so few lions in an area where hundreds of zebra graze?

The energy pyramid is a useful model that illustrates an important ecological idea: As food energy passes from level to level in a food chain, *the amount of energy available to the next-higher level decreases.*

Often the energy pyramid concept communicates the idea that as energy passes along a food chain, fewer and fewer organisms become involved. This is the case in the pond food chain illustration. It is also the case in most other food chain examples. There are exceptions to this idea, however. A rose bush, which is a single organism, can support a population of thousands of aphids (plant lice), which use the plant's fluids as food. A single oak tree may supply the food for hundreds of caterpillars, which devour its leaves. In both of these examples the numbers of organisms increase from producer to first-order consumer level, even though the amount of available food energy decreases.

If people wish to obtain the most food energy they can from cultivated plant crops, should they feed the grain to beef cattle and then consume the steak, or should they eat the plant products directly? Understanding the concept of food energy loss within a food chain should give you the answer to this question. Unfortunately, people often choose to disregard this knowledge.

An Energy Pyramid for a Pond Food Chain

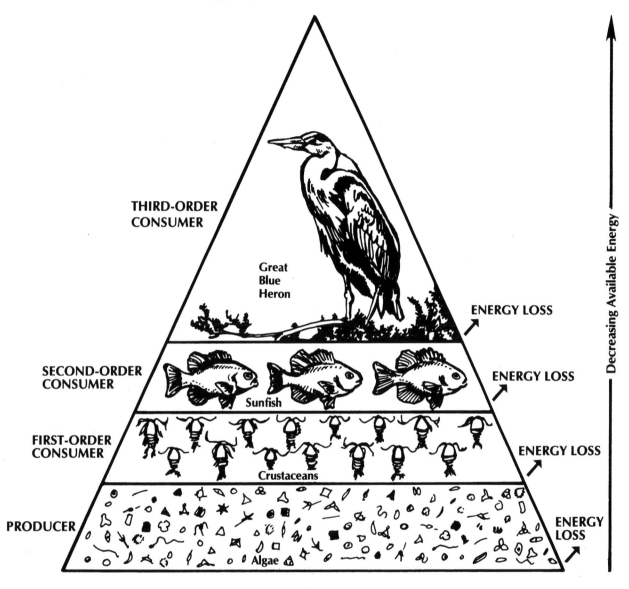

One Pond Food Chain

We have seen the tremendous importance of energy relationships in communities. Food chains and food webs are evidence of this. Food-energy relationships are often referred to as eater-eaten relationships. Why would this be the case?

There are other relationships of importance in living communities. Some of these involve food energy and some do not. Let's look at some of them.

Relationships in Communities

This caterpillar is being parasitized by the larvae of a tiny wasp. (Courtesy of H. R. Hungerford)

PARASITISM

Parasitism is one kind of community relationship. Although we do not particularly like **parasites** such as tapeworms, roundworms, fleas, lice, and mites, we have to admit that they have developed a very successful way of existence. A parasite is an organism that lives on or in another organism and obtains its energy from that organism **(host).** Animals like frogs, turtles, fish, foxes, cattle, and birds may have several different kinds of parasites living on or in them at the same time. It is even possible for one's pet dog to have mites, fleas, ticks, tapeworms, and other parasites all at the same time! This does the dog no good at all, but the parasites evidently find the environment provided by the dog to be very hospitable.

Even plants have parasites; mistletoe, corn smut, wheat rust, and apple rust are good examples. What are other examples of parasitic plants or animals? Is parasitism an example of an eater-eaten relationship? Explain your answer. What happens to the parasite if the host dies?

_____
_____
_____

SCAVENGING

Another kind of community relationship is **scavenging.** A **scavenger** is an organism that will eat another dead organism it has not killed. The vulture is a good example of a scavenger. The term *scavenger* is a tricky one because most organisms that get their food this way do so only part-time. Even the vulture, for example, will eat live food once in a while. Therefore, we might say that organisms that are scavengers are usually only part-time scavengers. A bear, for example, will eat dead animals on occasion. So will a fox, a wolf, a pet dog, or an eagle. It would seem, then, that most scavengers are also part-time predators.

Why is scavenging an example of an eater-eaten relationship? _____
_____
_____
_____

Turkey vultures. (Corel CD)

Relationships in Communities

COMMENSALISM

Commensalism is another relationship in communities. Commensalism is *not* an eater-eaten relationship. It is a relationship between two organisms in which one is benefited and the other is not affected in one way or another. For example, in Florida cypress swamps you can see "air plants," or bromeliads, growing from the bark of cypress trees, and orchids taking root on the trunks and branches of swamp trees. The bromeliads and the orchids benefit from the trees. Living above the surface of the swamp water, they get sunlight and perhaps some necessary chemicals from decaying debris that collects in the bark and around the tree roots. In any event, the tree is unaffected. It seems to survive just as well with the bromeliads and orchids as without them. The photograph below vividly illustrates commensalism.

You might find it interesting to investigate other examples of commensalism. The remora fish and the shark have a commensal relationship. So have bluebirds and woodpeckers. Can you think of another example?

_____
_____
_____

The bromeliads (center) and Spanish moss (like streamers hanging down) are two examples of commensalism. (Courtesy of H. R. Hungerford)

MUTUALISM

Mutualism is another fascinating relationship in nature. Like commensalism, mutualism involves two organisms, but in mutualism, *both* organisms benefit from the relationship.

There are many examples of mutualism in nature, of which the most classic is that of the yucca and yucca moth. The yucca moth is a small white insect that emerges from the pupa stage when the yucca plant is in bloom. The female yucca moth is the only insect that can pollinate the yucca plant. She does so, then she lays her eggs in the flower. The flower produces seeds, and the moth larvae will feed on some of these seeds. In this way, both organisms benefit and survive. In fact, either without the other would probably be doomed to extinction.

Lichens exhibit mutualism also. Lichens are small, often flat, green or greenish, found growing on rock surfaces or tree trunks. The lichen is a combination of two organisms: fungus living with a photosynthetic alga. The alga is the producer, providing food for the lichen. The lichen in turn provides protection for the alga and helps keep it from drying out. Both organisms benefit from the relationship.

Can you think of any other examples of mutualism? _____

Moss and lichen. (Corel CD)

Relationships in Communities

The mushroom growing here on the ground of a Wisconsin forest is an example of a decomposer.
(Courtesy of H. R. Hungerford)

DECOMPOSERS

A **decomposer** helps break down dead organisms and waste material produced in the community. Did you ever wonder what happens to all the leaves that fall from a forest's trees in autumn? What happens to the tons of sticks, branches, and animal wastes that might pile up in a community? These materials are broken down by decomposers in the community. Many organisms such as bacteria, molds, and fungi digest both plant and animal waste in order to obtain energy for themselves. And, while doing so, they release chemicals that are beneficial to green, growing plants.

Why are decomposers important parts of any food web? What role or roles do decomposers play in an eater-eaten relationship? _____
_____
_____

COMPETITION

Competition is a critical factor for organisms in any community. Animals and plants must compete successfully in the community to stay alive. It is not surprising to learn that in the prairie community shown on pages 31–32, for example, grass plants have to compete with each other for space, water, and soil nutrients. Or that the prairie dogs compete with each other for food, space, and mates, as do the buffalo. Here animals within the same species are competing with each other. This is called *intraspecies competition*. You might note a different example of intraspecies competition with the wrens nesting in your backyard. Each pair of wrens has a given territory. Any new wrens trying to move into that territory are driven off vigorously. This intraspecies competition of the wrens may have survival value for them.

If they didn't have a specific territory staked out, what might happen to their food supply, or to the nesting sites and materials for nest building available in that territory? What might happen to the baby wrens if ten pairs of wrens nested in the territory now set up and protected by one pair? _____

_____
_____
_____
_____
_____

The kind of competition that exists between plants and animals of different species is probably more familiar to you. Using the prairie again as an example, we can see that the grasses must compete with the milkweed as well as all other rooted green plants that can grow where the grass grows. Similarly, the prairie dogs compete with the buffalo and the elk for green plants to eat. This is known as *interspecies competition*.

Both kinds of competition are important influences in the community. While ecologists don't know all the facts of competition as yet, they are certain that intraspecies competition helps control population density. Why would this be important?

Cite three examples of interspecies competition among people. How does competition of both kinds affect the homeostatic balance of a community? _____

_____
_____
_____
_____
_____
_____

Relationships in Communities

The spacing between these ponderosa pine trees may very well be the result of intraspecies competition for sunlight and soil nutrients. (Courtesy of H. R. Hungerford)

Relationships in Communities

ACTIVITY 13 | Constructing a Food Web Showing Several Eater-Eaten Relationships

Choose any natural community you wish. Draw a food web below for that community that includes the following features: producer, first-order consumer, predator, parasite, scavenger, decomposer. Be certain that your food web is one you can defend logically. Label each step in the web and use arrows between organisms to show how they relate to each other.

Relationships in Communities

ACTIVITY 14 | Investigating Special Relationships

Some of the relationships that occur in communities are most interesting and unique. Use reference books to determine the specific relationships that exist in the lifestyles of each of the following organisms. Are they examples of mutualism, parasitism, commensalism, predator-**prey,** or decomposition? How does each influence the nature of its community? You might want to report your findings to your classmates.

Spanish moss _____

Indian pipe _____

dodder _____

yucca plant and yucca moth _____

shark and remora _____

bluebirds and woodpeckers _____

rhinoceros and tickbird _____

Urechis worm _____

hermit crab _____

ants and aphids _____

termites and flagellates _____

cowbirds _____

praying mantis _____

lichens _____

legumes and bacteria _____

Chapter 5 Review | Relationships in Communities

1. How does a food web differ from a food chain? _____
   _____
   _____

2. Why are there so few elephants in an ecosystem compared with the number of plants in that ecosystem? _____
   _____
   _____

3. How does mutualism differ from commensalism? _____
   _____
   _____

4. In what way are lichens unusual? _____
   _____
   _____

5. What does the term *food niche* mean? _____
   _____
   _____

NUTRIENT CYCLES | 6

Is it possible that one of the hydrogen atoms now inside your body was once in the body of George Washington or in a maple leaf? The answer is yes. It is possible because hydrogen is one of the **nutrient atoms** that are necessary for all organisms to live. There are approximately 30 of these atoms in all. Other important nutrient atoms are carbon, oxygen, nitrogen, phosphorus, and potassium. These atoms travel between the abiotic environment (air, oceans, rocks, soil, etc.) and the biotic matter (organisms) in food webs. Their paths are called **nutrient cycles.** Nutrient atoms are used over and over again. This is essential, because very little new matter from outer space lands on the earth. Each year, meteorites and cosmic dust enter the earth's atmosphere and fall to its surface, adding to the matter already here. The amount added is tiny when compared with the matter that has been here since our planet was formed. Since practically no new matter is added to the earth, we are living in a **closed ecosystem.** However, energy from the sun comes to the earth all the time.

In the last chapter, we discussed the energy in various food chains. Now we are going to focus on the *movement of chemicals* through a food chain. Let's use a simple food chain as an illustration. Grass takes in water (**inorganic** or abiotic material) from the soil. It also takes in carbon dioxide (another inorganic substance) from the air. The grass plants, using the sun's energy and the process of photosynthesis, chemically combine these inorganic compounds into sugar (an **organic** material). Plants can use the sugar for energy, or they can transform it into other organic compounds such as proteins, starch, or **lipids.** When the grass is eaten by a cow, those organic compounds (**nutrients**) are transferred to the cow. If you were to drink milk or eat a burger made from that cow, those nutrients would be transferred to you. As you move through your life cycle, going through the processes of respiration, excretion, and eventually death, those nutrient atoms will be returned to the atmosphere, the oceans, or the soil of the ecosystem. The decomposers will then reduce those nutrients back to inorganic atoms and compounds to be used again by other plants.

The food chain is really a nutrient cycle; in fact, it is part of the **carbon cycle.**

Nutrient Cycles

THE CARBON CYCLE

If you read Chapter 9 (Global Warming) before reading this one, you have already seen a carbon cycle diagram. In this diagram, the focus is on the cycling of carbon between the physical environment and living organisms. **Carbon** atoms are very important to us on Earth. They are found in sugars, starches, proteins, lipids, and other biological molecules. Carbon atoms are involved in all processes necessary for life. Examine the diagram below carefully.

THE CARBON CYCLE

Organic sediments are formed when dead ocean animals and plants decay. The process of decay is carried out by decomposers (fungi, worms, and bacteria).

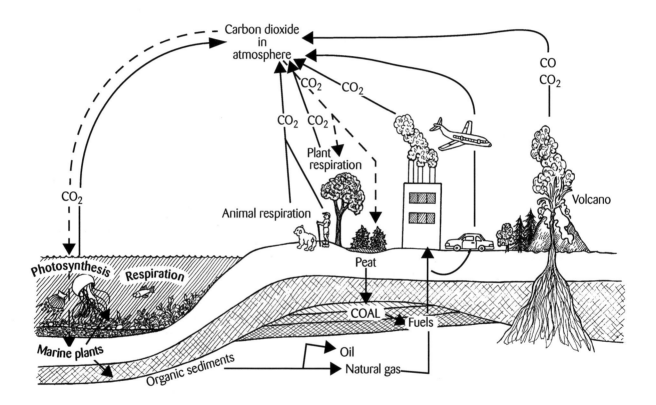

Nutrient Cycles

1. Using the carbon cycle diagram on page 70, list three carbon dioxide sources.
   _____
   _____
   _____

2. List three carbon sinks.
   _____
   _____
   _____

3. If there were no energy supplied by the sun, the carbon cycle would stop. Why?
   _____
   _____
   _____

4. Trace a carbon atom from a marine plant into the atmosphere and then back again. There are two possible paths. Trace both paths.

   Path 1 _____
   _____

   Path 2 _____
   _____

5. If plants were removed from the ecosystem, could carbon atoms recycle? _____
   Explain your answer. _____
   _____
   _____

6. If humans and animals were removed from the ecosystem, could carbon atoms recycle?
   _____ Explain. _____
   _____
   _____

Nutrient Cycles

THE WATER CYCLE

Water is essential to all forms of life. It can take more than two weeks for a person to die of starvation. However, two days without water can result in death. Even bones and teeth, which most people think of as hard and dry, contain about 5 percent water. Study the water cycle diagram below carefully, then answer the questions that follow.

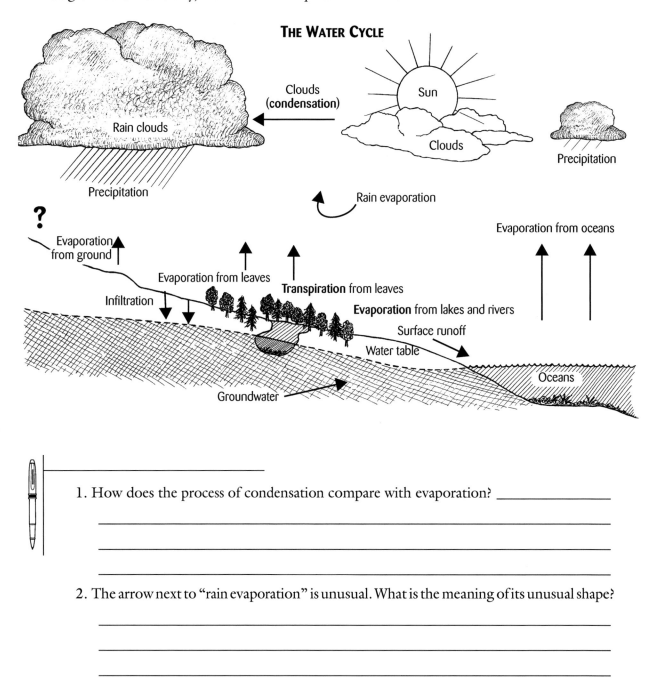

1. How does the process of condensation compare with evaporation? _____

_____

_____

_____

2. The arrow next to "rain evaporation" is unusual. What is the meaning of its unusual shape?

_____

_____

_____

Understanding Basic Ecological Concepts

3. Locate the "water table." Can you figure out what it is by looking at the diagram? (Take a guess if you are not sure of its meaning.) _____
_____
_____
_____

4. Do the same for the term "infiltration." _____
_____
_____
_____

5. There is something missing from this picture. It is indicated by the question mark at the far left of the diagram. There is room for you to draw in the missing item(s) and the related arrows. If you think you know what is missing, *check with your teacher first* before you draw.

ACTIVITY 15 | A Do-It-Yourself Water Cycle

Go to the chalkboard, or, if you are at home, stand in front of a mirror. Get close, no more than 10 cm away.

Face the chalkboard or mirror. Take in a deep breath, then exhale completely—as quickly as you can—onto the surface of the board or mirror. Step back. What do you observe?

_____
_____

How does that relate to the water-cycle diagram? _____
_____
_____

Nutrient Cycles

THE PHOSPHORUS CYCLE

Phosphorus is another important nutrient that is involved in the transfer of energy in cells. It is one of the elements needed to make DNA and RNA. The decay organisms in the phosphorus cycle are the same as those mentioned in the carbon cycle and water cycle previously discussed. These decay organisms include fungi, worms, insects, and bacteria.

Examine the phosphorus cycle diagram below. Notice that soil absorbs and holds phosphorus much better than the oceans or fresh water do.

THE PHOSPHORUS CYCLE

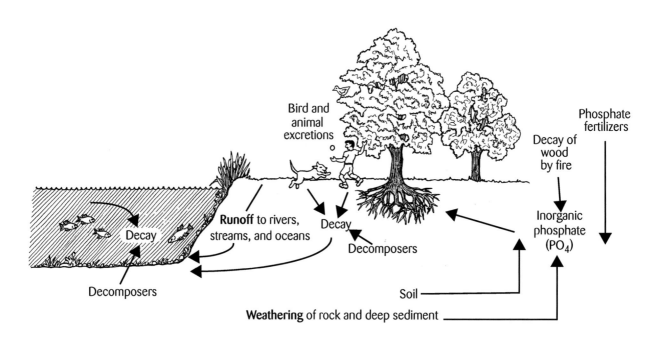

1. What is the main sink (reservoir) for phosphorus? _____

2. How do animals get their phosphorus? _____

3. How much phosphorus can be found in the air? _____

4. What does the term *runoff* mean? _____

Nutrient Cycles

There are other nutrient cycles in nature that we will not discuss in detail. Two examples are the potassium cycle and the complicated nitrogen cycle. Bear in mind that all of these cycles interact with each other.

As discussed at the beginning of this chapter, we consider the planet Earth to be a closed ecosystem, with all of these cycles going on at the same time. We separate the cycles and study them one at a time just to make it easier for us. Now, look back at the three cycles we have studied, then answer the questions below.

1. What things can be found in all three cycles? _____

_____

_____

2. What organisms are involved in decay and decomposition? _____

_____

_____

3. What chemicals do green plants assimilate through their roots as shown in the three cycles? _____

_____

_____

4. What chemicals do green plants assimilate through their leaves? _____

_____

_____

Nutrient Cycles

ACTIVITY 16 | Creating a Closed Ecosystem

In this activity you are going to prepare and observe a closed ecosystem similar to the one in which you are living. However, the one you set up will be on a very small scale!

Materials	
Large culture tube with a screw cap (200 mm in length) or Clear glass jar with a screw cap (needs a rubber seal beneath cap) Pond water snail Wide-spectrum growing lamp and fixture	Aquarium gravel/sand Electric light timer Spring water/pond water A healthy sprig of anacharis (*Elodea*) *Optional: A sealed, self-sustaining ecosystem (obtained from a scientific supply house)

Procedure

1. Add the sand and gravel to a depth of 2 cm in the test tube or jar.

2. Fill the container with spring or pond water, leaving the top 3 cm empty.

3. Add the healthy sprig of anacharis (you can anchor it in the sand or let it float freely).

4. Add the snail. Now cover the tube or jar as tightly as you can. *Be careful that you don't break the test tube or jar.*

5. Place the setup in a location where it will *not* receive direct sunlight. Place the lamp so that it is approximately 6 cm away from the container. The light should shine on a glass side of the container. You do *not* want the light to shine on the top cover!

6. Set the timer so that the light source will be on from 6:00 A.M. to 9:00 P.M. Connect the light source to the timer and the timer to an electrical circuit.

7. Predict how long the snail will live without your adding food to the setup.

   _____

8. Leave the set undisturbed for 30 days. Examine on a daily basis, checking on the health of the snail and the plant. Record your observations on a data table (see the sample below). If the plant or the snail appears to be sickly, stop the activity. Place the snail and plant in an aquarium so that they can recover.

Nutrient Cycles

DATA TABLE*

Day	Health of Snail (appearance, movements, etc.)	Health of Plant (color changes, number of new leaves formed, etc.)
1	S	
2	A	
3	M	
4	P	
5	L	
6	E	

*Prepare your data table on a piece of paper or on the computer, and have space for 30 days of observations.

Questions

1. After two days, examine the top 10 cm of the tube or jar for evidence of condensation. What evidence did you find? _____

2. Why were you advised *not* to place the closed ecosystem in direct sunlight? _____

3. Which of the three cycles we have examined are going on continually in the mini-closed ecosystem? _____

   Explain your reasoning for each cycle. _____

4. How does the plant help keep the snail alive? _____

5. How does the snail help keep the plant alive? _____

6. What decomposers are present in your ecosystem? _____

Chapter 6 Review | Nutrient Cycles

1. Why is the cycling of carbon atoms essential to living things? _____
   _____
   _____

2. What role does the sun play in the carbon cycle? _____
   _____
   _____

3. What role does the sun play in the water cycle? _____
   _____
   _____

4. Explain why atoms have to be recycled in order for life to continue on Earth.
   _____
   _____
   _____

5. What role do the decomposers play in the cycles studied? _____
   _____
   _____

SUCCESSION 7

Sometimes people coin a phrase that catches on, even though it might not be completely accurate. This is true of the term *balanced community*. *Balance* suggests stability—that there are no changes taking place. As you know, this just doesn't happen in natural communities. A community is a dynamic, active, changing environment, with animals and plants interacting with one another all the time. There are long-term changes and short-term changes. Populations fluctuate. Consider the effect of bird migration on a forest in the Midwest. On May 10 there may be hundreds of Wilson's warblers flitting about in the treetops. They may remain through the day and be gone on May 11, headed north toward their breeding grounds in Minnesota and Canada. The effect of hundreds of Wilson's warblers acting as second-order consumers in the forest for an entire day may be substantial. Still, they are here today and gone tomorrow. There is nothing **static** about this situation.

What should we consider a balanced community to be? Is the forest through which hundreds of Wilson's warblers pass considered balanced? The answer is yes. If the forest retains its basic characteristics for fairly long periods of time, it can be called a balanced, homeostatic community. In reality, any self-maintaining community is balanced. The fact that Wilson's warblers visit the forest for brief periods has little to do with the long-term condition of the forest.

Often ecologists like to refer to a homeostatic community as one where changes do not destroy the character of the community. A grassland community, for example, remains a grassland community although the numbers of grazing animals may fluctuate dramatically. An oak-hickory forest remains an oak-hickory forest even though dramatic species differences can be noted from time to time. The ability of a community to retain its basic characteristics is the best explanation of a "balanced, homeostatic community."

A CLIMAX COMMUNITY

The grassland community and the oak-hickory forest may retain their basic character for hundreds of years. Of course, if the grasslands are plowed and planted to grow wheat, the situation changes drastically. So, too, with the forest if it is cut or if fire destroys it. Without such drastic changes, however, the grasslands and the oak-hickory forest (in some regions) have the ability to remain much as they have for centuries. Some communities are like this. They are dynamic environments, but their basic character remains unchanged.

Any community that can retain its character for many hundreds—or thousands—of years is called a **climax community.** The term *climax* refers here to the *last* community or the *most stable* community for a particular area. Other examples of a climax community would be the rain forest, the cactus desert, the **taiga,** and the **tundra.** You will be learning later about how climax communities develop. But first look at the illustration that follows.

HOW COMMUNITIES CHANGE

Communities can, and do, change. A particular **coniferous** forest may have remained stable over several hundred years, but it can catch fire from a single bolt of lightning or from a carelessly thrown cigarette or match and be completely destroyed.

If a forest is destroyed by fire or from lumbering, will it grow back as it was before? Will oaks and hickories grow in the ashes of a former oak-hickory forest? The answer to the first question is yes—eventually. The answer to the second question is no. Although this doesn't seem reasonable, it's a fact. If the soil of the forest is not completely destroyed by fire, for example, the first plants to appear will be soft-bodied plants such as grasses and flowering plants. These are eventually replaced by what are called **pioneer trees**—trees far different from the oaks and hickories that originally grew there. The pioneer trees provide an environment in which a new group of trees can eventually grow.

After many more years, oak and hickory seedlings will begin to be seen in the area again. However, it may take 200 years for an oak-hickory forest to return to its original state after a fire or lumbering has destroyed it.

The oak-hickory forest seen here is a climax community in the midwestern United States.
(Courtesy of H. R. Hungerford)

The replacement of one community by another is called **succession.** Succession is a very normal happening in nature and is evident in many places. Succession can be seen in an abandoned farm field. It can be seen at any pond, in any lake, and at any bog. It can be seen in a vacant city lot where plants and animals are mostly unmolested. Succession can even be seen where grasses and other plants grow in the cracks of sidewalks.

The interesting thing about succession is that it is usually an orderly, predictable process. Ecologists can predict what will happen in a given environment in 10 years, in 100 years, in 200 years. Almost all ponds and lakes, for example, follow a natural sequence of events that eventually end in the production of fairly dry land, along with a land community established where a pond or lake once existed. This may be hard to believe, but it is precisely the case.

It was noted earlier that a forest community can be completely destroyed by lumbering or fire. When this happens in a hilly area, it is quite possible that the tragedy will be followed by serious soil **erosion.** Solid rock is usually found under the layer of soil in a hilly region. If the soil is completely stripped off the land, it will be centuries before it can return to its original state. But given time—perhaps 1,000 years—the scars can be healed and a new forest can grow where only bare rock exists today.

This, too, may be hard to believe, but it is true. How does such a progression of events occur? We will call this process *bare rock succession,* which is, in fact, a surprisingly orderly series of events.

Bare rock succession begins with a rock outcrop such as the one shown here. (Courtesy of H. R. Hungerford)

Succession

As hilly rocks are laid bare by erosion, they are exposed to the atmosphere. The geologic processes of weathering can begin, and the rock is slowly broken down into smaller and smaller fragments. Many forms of **weathering** exist, and all are effective in reducing bare rock to smaller particles.

The first organisms to appear on the bare rock are lichens. These hardy organisms grow on the rock itself. They produce weak acids that assist in the slow weathering of the rock surface. The lichens also trap wind-blown soil particles. These eventually produce a very thin layer of soil—a change in environmental conditions that gives rise to the next stage in bare rock succession.

Mosses are able to grow in the soil provided by weathering and the lichens. They produce a larger growing area and trap even more soil particles. Mosses also provide a moister environment on the rocks. The combination of additional soil and moisture establishes abiotic conditions that favor the next successional stage.

The seeds of **herbaceous** plants now invade what was once only a bare rock surface. Grasses and other flowering plants take hold. Organic matter provided by the dead tissue of plant bodies is added to the thin layer of soil, while the rock continues to be weathered from below. More and more animals join the community as it becomes larger and more complex.

By this time the plant and animal community is a fairly complicated one. The next major invasion is by weedy shrubs, which are able to survive in the amount of soil and moisture provided. Time passes, and the process of building soil speeds up as more and more plants and animals invade the area. Soon trees take root, and forest succession is evident. Of course, many years will pass before a climax forest grows here, but the scene is set for that very happening.

Of great interest in bare rock succession is the fact that each stage in the pattern dooms the community that existed before it. Mosses provide a habitat most inhospitable to lichens, the herbs will eventually destroy the moss community, until the climax stage is reached.

Intraspecies and interspecies competition, and changes in other biotic and abiotic conditions, greatly influence succession stages. When the climax stage is reached, competition and other interactions still exist. However, a homeostatic balance exists so that the climax community creates environmental conditions favoring its existence rather than dooming it.

In summary, ecologists have found that the succession patterns of different communities usually have the following characteristics in common:
- A change in the plant and animal community members. In general, as community members change, there is an increase in the number of species present as well as in the complexity of the community structure.
- An increase in organic matter from stage to stage.
- A tendency toward greater homeostatic stability as stages progress.

Each example of natural succession has identifiable stages. These stages change slowly over the years. The stages can be predicted. And, in each stage of succession, the dominant community actually creates environmental changes that doom it. The exception to this is the final or climax stage, which is self-renewing.

Bare Rock to Forest Soil

A. Bare rock is exposed to the elements.

B. Rocks become colonized by lichens.

C. Mosses replace the lichens.

D. Grasses and flowering plants replace the mosses.

E. Woody shrubs begin replacing the grasses and flowering plants.

F. A forest eventually grows where bare rock once existed.

Hundreds of Years →

Succession

ACTIVITY 17 | Projecting Pond Succession

The photo below shows a small emergent pond in the Midwest. Both pond weed and cattails are emerging from its surface. The water itself is barely visible. This community is an extremely rich one. From your knowledge of succession, provide the best possible answers to the following questions.

Pond succession. (Courtesy of H. R. Hungerford)

1. Is this pond young, middle-aged, or old? What observations led to your conclusions?

_____

_____

_____

2. What will be the appearance of this pond in another 10 years? _____

In 100 years? _____

3. How are the animal species in this present community likely to change as the years go by? Use specific examples if possible—such as water snakes, frogs, salamanders, water insects, snails, red-winged blackbirds, and herons. _____

4. How could pond succession be stopped, at least temporarily? What is a second way that this might be accomplished? Mention a third way, if you can.

Succession

ACTIVITY 18 | Studying Forest Succession

You already know what succession is. How do ecologists study succession? And what do their studies tell them?

Ecologists collect data about succession in different communities in different ways. As you know, one of the techniques they use on prairies and in forests is called the **quadrat.** Let's investigate this technique in more detail.

A quadrat is nothing more than four stakes with twine or clothesline set out in a square or rectangle of exact dimensions. Ecologists collect data inside the quadrat. On a prairie, where plants grow very close together, ecologists may set up a quadrat enclosing only one square meter. In a forest, they may find it more useful to set up quadrats of 100 square meters (100 m^2). If they are studying shrubs or herbs, they may use rectangles of 10 square meters. Sometimes scientists will set up both kinds of quadrats in the same area. The following drawing will give you an idea of how a quadrat is set up in a forest. The **canopy** consists of the trees that form the uppermost branchy layer of the forest. The **understory** includes all other lower trees and shrubs.

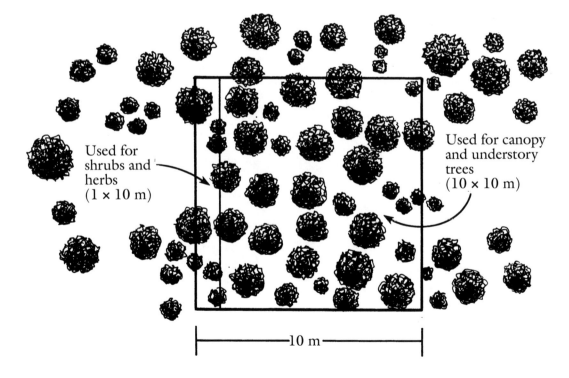

© 1979, 1989, 2001 J. Weston Walch, Publisher

Ecologists use a quadrat map or worksheet to record exactly where the trees in the quadrat are and what kinds there are. From this record they can work out the total populations that exist in the quadrat. From this information, ecologists can draw conclusions about the forest's characteristics in general. They may study several quadrats and use averages to help them obtain the most accurate information about the makeup of forest populations. Figure AO—A Model (on page 88) is an example of one forest quadrat. Study this model.

In the Figure AO model, the ecologists were studying the canopy trees in a forest 100 years old. Each tree species has its own symbol. The records show that there were six trees (made up of three species) in the canopy of this quadrat. These tree species are typical of trees found in a healthy older forest in southern Indiana or Illinois. Use the records in Figure AO to calculate the population densities of the canopy trees. Do this before going on.

_____

_____

Directions for Completing Activity 18

On the pages following Figure AO, you will find a series of five other figures showing five different quadrats. Each quadrat is an example of what you might find in a forest in southern Indiana (or Illinois) if you studied that forest over a period of 45 years. The first quadrat (Figure AA) is a seedling count in a field after it has been taken out of cultivation for 5 years. The next two quadrats (Figures AB and AC) are canopy and understory quadrats 20 years after cultivation stopped. The last two quadrats (Figures AD and AE) are canopy and understory quadrats 50 years after the field was abandoned. With three of these quadrats you will find air and soil temperature data.

Study each quadrat. Figure out and record the total populations in each. Also compute the percentages for each population, as was done in the model (Figure AO). When you have collected all the data, you should be able to draw some important conclusions about this kind of succession.

Succession

FIGURE AO—A MODEL
Tree Species Distribution in a 100 m² Quadrat Field in Southern Indiana, Abandoned 100 Years

Canopy Species:

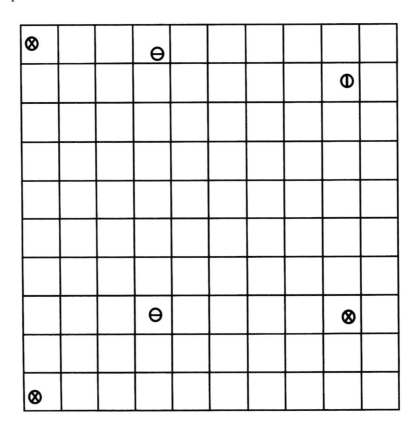

	How Many	Percent of Total (%)
Key: Shagbark hickory ⓘ	1	16.67
Yellow chestnut oak ⊖	2	33.33
Red oak ⊗	3	50.00
Total	6	100.00

Succession

Note: The numbers in the table for Figure AO were determined by counting the number of each species in the quadrat. This number is recorded in the blank to the right of each species' name. When this column is complete, the total number is obtained by adding this column. A percentage of the total for each species is then calculated. These percentages are added together to check the calculation.

FIGURE AA
Tree Species Distribution in a 100 m² Quadrat
Field in Southern Indiana, Abandoned 5 Years

Seedlings Present:

	How Many	**Percent of Total (%)**
Key: Sassafras △	_____	_____
Persimmon ●	_____	_____
Total	_____	_____

June 15—11:00 A.M. — Sunny Day
Air temperature — 27 degrees C (80 °F)
Soil surface temperature — 35 degrees C (95 °F)

© 1979, 1989, 2001 J. Weston Walch, Publisher   89   *Understanding Basic Ecological Concepts*

Succession

Figure AB
Tree Species Distribution in a 100 m² Quadrat
Field in Southern Indiana, Abandoned 20 Years

Canopy Species:

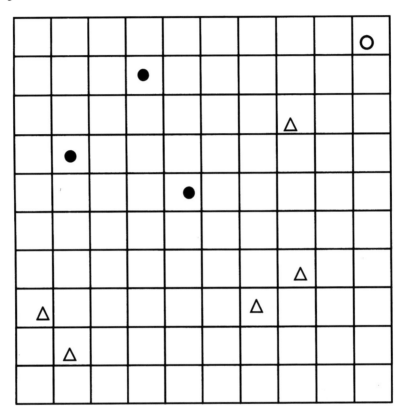

	How Many	**Percent of Total (%)**
Key: Winged elm ○	_____	_____
Persimmon ●	_____	_____
Sassafras △	_____	_____
Total	_____	_____

June 15—11:00 A.M. — Sunny Day
Air temperature — 24 degrees C (75 °F)
Soil surface temperature — 29 degrees C (84 °F)

© 1979, 1989, 2001 J. Weston Walch, Publisher

Understanding Basic Ecological Concepts

Succession

Figure AC
Tree Species Distribution in a 100 m² Quadrat Field in Southern Indiana, Abandoned 20 Years

Understory Species:

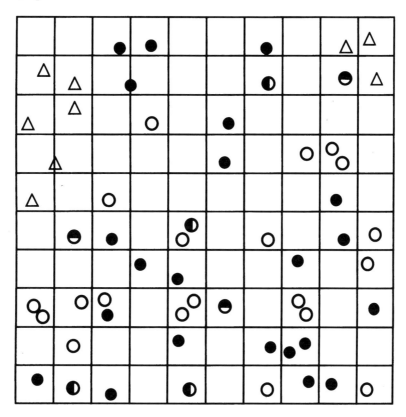

	How Many	**Percent of Total (%)**
Key: Winged elm ○	_____	_____
Persimmon ●	_____	_____
Sassafras △	_____	_____
Black cherry ⊖	_____	_____
White ash ◐	_____	_____
Total	_____	_____

© 1979, 1989, 2001 J. Weston Walch, Publisher

Understanding Basic Ecological Concepts

Succession

Figure AD
Tree Species Distribution in a 100 m² Quadrat
Field in Southern Indiana, Abandoned 50 Years

Canopy Species:

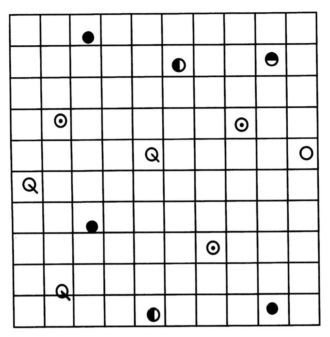

	How Many	Percent of Total (%)

Key: Black Cherry ◐

White Ash ◐

Persimmon ●

Winged Elm ○

Bitternut Hickory ⊙

White Oak Q

Total

June 15—10:00 A.M. — Sunny Day
Air temperature — 21 degrees C (70 °F)
Soil surface temperature — 20 degrees C (68 °F)

Figure AE
Tree Species Distribution in a 100 m² Quadrat Field in Southern Indiana, Abandoned 50 Years

Understory Species:

⊗				⊖					
								⊕	
		⊙				⊖			
			⊙					⊙	⊙
Q						Q			
	Q								
	⊕			⊖				⊗	
		Q							
⊗			Q					Q	

	How Many	Percent of Total (%)
Key: Bitternut hickory ⊙	_____	_____
Shagbark hickory ⊕	_____	_____
White oak Q	_____	_____
Yellow chestnut oak ⊖	_____	_____
Red oak ⊗	_____	_____
Total	_____	_____

Succession

Specific Tasks for This Problem

1. The 5-year quadrat you studied (Figure AA) represents a part of a field where cultivation was discontinued only 5 years ago. Hypothesize why so few seedlings appear in this quadrat. _____
   _____
   _____
   _____
   _____

2. Compare the understory quadrats (Figures AC and AE) and the canopy quadrats (Figures AB and AD). Place a check in the appropriate space below indicating whether there are more trees in the understory or in the canopy at any given time.
   understory _____ canopy _____
   How can you explain this observation? _____
   _____
   _____
   _____
   _____

3. Using the quadrat figures, provide the following information:

   (a) List the tree species that disappeared from the understory between 20 and 50 years.
   _____
   _____
   _____

   (b) List the tree species that appeared in the understory between 20 and 50 years.
   _____
   _____
   _____

   (c) Compare these two lists and note the differences below. _____
   _____
   _____
   _____
   _____

Succession

4. List the tree species that are found in the canopy at 50 years. _____
   _____
   _____
   _____
   _____

   Using the 50-year understory data (Figure AE), predict which of the trees you listed above will not be in the canopy at 100 years. What is the basis for your prediction?

   Prediction: _____
   _____
   _____

   Reason: _____
   _____
   _____

5. There is a difference of 6 °C air temperature and of 15 °C in soil surface temperature between the 5-year quadrat and the 50-year quadrat. What might account for this difference? _____
   _____
   _____
   _____
   _____

6. A climax forest is a forest in which succession has, for the most part, stopped. One kind of climax forest is called an oak-hickory climax, because oaks and hickories are the predominant trees in it. Another kind is a beech-maple climax forest. Using Figures AD and AE, predict what kind of climax forest will be produced here. What are the reasons for your **prediction?**

   Predicted climax forest type: _____
   _____

   Reason: _____
   _____
   _____

Succession

7. From the data you have gathered concerning the various quadrats, predict whether persimmon or sassafras trees would ever be commonly found in a climax forest. Give the reasons for your prediction.

   Prediction: _____

   Reason: _____

   _____

   _____

8. Are there fewer or more trees in the canopy of Figure AO than the canopy of Figure AD? Check the appropriate answer (please note the age of each canopy):

   More in AO _____   Less in AO _____

   Hypothesize why this might be the case. _____

   _____

   _____

   _____

9. After studying the data you have (including Figure AO), list below the events (changes) that take place during succession in a southern Indiana forest studied from the time that it was abandoned until it is 100 years old.

   _____

   _____

   _____

   _____

   _____

   _____

   _____

   _____

   _____

ACTIVITY 19 | Studying Succession on Your Own

Now that you have completed a paper-and-pencil laboratory on succession, you should have enough experience to study this phenomenon in the real world.

Choose a natural community in your immediate vicinity. You might find two wooded areas of different ages, or a young pond and an older one, or two vacant city lots of different ages. If you live in an area where road construction is taking place, you can compare the recently bulldozed roadsides with those that have not been touched. The fence rows between cultivated farms would also be ideal for comparison with the fields themselves.

As you observe these two areas, compare them in terms of the kinds and numbers of organisms present. Which has a greater species diversity? How do the abiotic conditions in each differ? What evidence can you find that supports the idea that succession is taking place?

Your teacher may even assist you in making quadrat studies of the two areas. This type of quantitative analysis should help you better understand the complex process called succession.

Succession

Chapter 7 Review | Nutrient Cycles

1. Why do communities change over time? _____
   _____
   _____

2. How are forest communities destroyed? _____
   _____
   _____

3. Why might the destruction of a climax community lead to loss of its topsoil?
   _____
   _____
   _____

4. Why are lichens called "pioneers?" _____
   _____
   _____

5. Why is the term *balanced* a poor one to use to describe a climax community?
   _____
   _____
   _____

HUMAN ECOLOGY 8

The number of people on the earth is increasing by more than 90 million individuals each year. These new citizens and those of us already here need places in which to live and work, as well as many services. So, through the use of science and **technology,** we are building homes, factories, electrical power plants, hospitals, offices, stores, shopping centers, roads, waste treatment plants, and much else in order to meet our many needs. The graph that follows shows the dramatic increase in our world's population since the Industrial Revolution.

As we strive to improve the quality of life throughout the entire world, we change ecosystems and destroy ecological communities. The results, more often than not, spell ecological disaster.

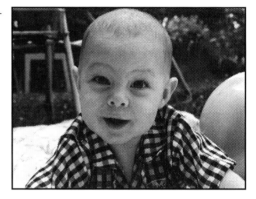

CARRYING CAPACITY

After reading the paragraphs above, you might conclude that **overpopulation** is the cause of most of our ecological problems. As you will see, a better term to use is **carrying capacity,** because there is no set amount of people supportable by the environment. In fact, a modern city can support many more people than cities of earlier centuries.

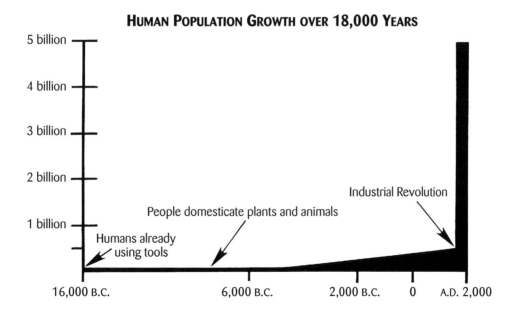

HUMAN POPULATION GROWTH OVER 18,000 YEARS

Human Ecology

Human carrying capacity can be defined as "the number of people that can be supported by the environment." The carrying capacity is affected by space and other factors. For example, in Mexico, on that country's border with the United States, many towns called *shantytowns* were created at the end of the twentieth century. The houses are made of cardboard and are crowded together. These shantytowns are still growing in the twenty-first century. One million Mexicans immigrated to this border area from the Mexican interior. They come *not* to cross into the United States, but to work in factories in Mexico. Starting in about 1965, many large American companies such as General Electric have built factories and industrial plants on the Mexican side of the border. Labor costs are much cheaper than in the United States. Mexican taxes are also much lower than in the United States.

The air, water, and soil in these areas are polluted with human and industrial wastes. Owing to the large number of people crowded into a small area, only 10 percent of the people can get clean water. Sixty-five percent of the sewage produced daily goes untreated, because most of the shacks are not connected to the public sewage system.

On the positive side, the factory executives have developed recycling programs and safety measures. They have also built hospitals and medical clinics for the workers. Clearly, however, these modern-day shantytowns do not have the capacity to support the huge number of people living in them.

List the factors other than space that affect human carrying capacity.
_____
_____
_____
_____
_____

Your list should include food, fresh water, air, topsoil, climate, and the ability to dispose of wastes. As you can see, space is not the main factor.

ATMOSPHERIC QUALITY

Food supplies depend on topsoil as well as on adequate amounts of water and minerals and on favorable temperatures. Erosion and the building of roads, parking lots, shopping centers, and structures of all kinds reduce the amount of topsoil available for growing food each year.

Fresh water used to be thought of as **renewable.** However, water is actually a **limited resource** that must be used again and again, as evidenced in the water cycle. Since it is a limited resource, our water's purity must be protected. Moreover, our water supply becomes even more limited when we take more of it out of the ground than is returned. Paving the land means that water from the clouds runs off into a sewer rather than percolating through the soil. Normally, water seeps through layers of soil and ends up in underground storage spaces called **aquifers.** More ground water is lost by being made unfit for human use when **toxic** wastes are dumped on the ground through which the water passes on its way to the aquifers.

The quality of our atmosphere is a critical factor that affects the carrying capacity of our planet. Serious problems associated with the atmosphere are **regional haze, smog, acid deposition,** a thinning of the **ozone layer,** and **global warming.** Since air currents circle the globe, what is added to the atmosphere in a distant land may be transported around the world in a short time. Human activities in one place can affect the environment in distant locations. An example of the movement of air pollutants over long distances in the atmosphere is **acid precipitation.**

Air pollution is everyone's problem. Adults breathe about 13,000 liters of air daily. Children—who are more active than adults, and whose lungs are still growing—are at greater risk from air pollution than adults. Older adults with heart or breathing problems also suffer needlessly due to air pollution. Cars, buses, trucks, and sport utility vehicles are responsible for one fourth of the total energy used in the United States. These vehicles also release harmful chemicals into the air. The other 75 percent of energy used comes from industrial sources and power plants. These facts should be kept in mind when you read this chapter and the next.

REGIONAL HAZE

If you have ever had trouble looking at a distant building because the air looked "dirty," you have witnessed a case of regional haze. Tiny particles, formed by sulfur dioxide reacting with other chemicals in the air, are mainly responsible for this type of pollution. Power plants are the principal source of sulfur dioxide. Also responsible for regional haze, but to a lesser degree, are nitrate particles formed from nitrogen oxides. Sulfur dioxide and nitrogen oxides enter our atmosphere when fuels in automobiles, power plants, and factories are burned for energy.

The United States government plans to have electric power plants lower their release of sulfur dioxide by 50 percent (over 1980 levels) by the year 2010. That would reduce regional haze by 30 percent. However, this goal may be reduced because of the increased need for electric power throughout the country. In 2001, for example, the shortage of electric power in California created *rolling blackouts*. Other parts of the country also were concerned about possible electrical shortages.

SMOG

In the summertime—or during very warm sunny weather—a brownish haze often pollutes our air. This is **smog.** Smog makes it difficult for people to breathe; it also reduces visibility. Most smog comes from the **emissions** produced by burning petroleum products or coal. The worst of these emissions are nitrogen oxides. Most of these gases are released in the exhaust of automobiles. Electric power plants also add a smaller amount of nitrogen oxide gases. In strong sunlight, nitrogen oxides react with oxygen and other chemicals in the air to produce **ozone gas.** Ozone of this kind stays close to the earth and does not join the ozone layer high up in the **stratosphere.** When close to the surface of the earth, ozone is harmful, producing smog. Stratospheric ozone, on the other hand, is helpful. (Both smog and ozone will be discussed in later chapters.)

To get the power plants, factories, and truck and bus manufacturers to reduce the level of pollutants they release into the air, you and your friends and family members can write and talk to your state and federal governments representatives. However, there are things that you and your family can do on a regular basis to reduce air pollution. For example, instead of driving to the mall to shop, shop by phone or over the Internet or by regular mail.

List three additional things that you and your family could do to reduce air pollution.

_____
_____
_____

Acid Deposition

There are both **wet** and **dry acid deposition.** Each contributes about 50 percent of the acidity in the atmosphere.

Wet deposition refers to **acid rain and snow.** Sulfur dioxide and nitrogen oxides, the same troublesome gases responsible for regional haze, are mainly responsible for acid deposition. These gases react with water in the atmosphere, forming weak solutions of sulfuric acid and nitric acid. The acids travel within the air masses in which they were produced. When the clouds in these air masses release their moisture in the form of rain and snow, the acids fall into lakes, oceans, and forests. The result is the killing of fish, water plants, and trees.

Dry deposition refers to acid particles in the atmosphere including dust and **soot.** These acid particles fall back to earth and are carried by the wind. They land on cars, buildings, and trees. During a rainstorm, these acid particles are washed off, adding more acid to the acid rain water on the ground. This combination is worse than the acid rain alone.

All trees, but especially pine trees, are killed by acid deposition. The acid causes the trees to lose their needles or leaves; it also damages the bark. The damage is caused both by direct contact—through acid rain—and by toxins in the soil that accumulate when acid rain falls to the earth.

Lakes are damaged by acid rain. Fish and insects in the lakes die. Birds that eat dying fish or insects are also killed. Fish eggs are easily destroyed by acid water.

What happened recently in Scandinavia is a good example of the acid deposition problem. Scandinavia has large areas that are rural and uninhabited. Officials began to notice fewer and fewer fish and other aquatic organisms in the lakes in these regions. Eventually, the lakes became lifeless. Scientists soon discovered that these lakes were highly acidic. For a while, the cause was a mystery. Eventually, the acid condition was traced (through several different experiments) to factories in industrialized areas in England. The smoke from the factories was rich in sulfur dioxide and oxides of nitrogen. This smoke was carried more than 1,000 kilometers to Scandinavia by the prevailing winds.

Acid rain in the northeastern United States has affected fish levels in the lakes and has killed many trees in the forests. In the year 2000, the states of New York and Connecticut sued 17 power plants in the Midwest and the Virginias in order to force them to reduce their emissions that were causing acid rain and smog. Many lakes and trees in the forests of the Adirondack mountains are dead. A dead lake is one in which all of the fish and other animals have been killed. Connecticut forests are in the same terrible state as those in New York. This is because the winds and weather usually travel from west to east. The sulfur dioxide and nitrogen oxides emitted from power plants in the Midwest end up on the East Coast.

Human Ecology

ACTIVITY 20 | Where in the United States Did Acid Rain Fall in 1999?

In order to understand the map below, it is necessary to know how we measure acid rain. A measurement called the **pH** scale is used. *The lower the pH, the more acid the rain is.* A pH of 7 is neutral, that is, neither acidic nor basic. Normal rain is slightly acidic, having a pH around 5.5. In the year 1999, one state had acid rain with a pH of 4.2 recorded. This was the worst case.

National Atmospheric Deposition Program (NRSP-3)/National Trends Network (2000). NADP Program Office, Illinois State Water Survey, 2204 Griffith Dr., Champaign, IL 61820.

Use classroom maps or atlases if you need help in determining the names of the states so that you can answer the following questions.

- In which state did rain with a pH of 4.2 fall? _____
- Find the state of New York. What was its lowest pH? _____
- Locate Massachusetts. What was its lowest pH? _____
- Find your state. What was its lowest pH? _____
- How do pH readings for rain in the West and Midwest compare with that in the Northeast?
- Which state(s) had no acid rain (normal rain has a pH of 5.5)? _____

Human Ecology

An early proposal to solve local air pollution problems was to build taller smokestacks. Scientists reasoned that releasing the gases at a higher altitude would dilute them and spread them over a larger area. Unfortunately, acid precipitation was created which fell on distant forests and lakes.

The only sensible answer seems to be to limit the amounts of sulfur dioxide and oxides of nitrogen released into the atmosphere. The dangerous pollutants that are in coal can be removed by crushing the coal and then washing it. This is an expensive process, though, and power plants don't want to spend the money to do it. The government seems reluctant to force them. The United States Congress has been considering legislation to force power companies to reduce the acid rain gases their plants emit. Oil companies are starting to remove sulfur from gasoline. This will be helpful in reducing acid rain, but it is not a complete answer to the acid deposition problem.

ACTIVITY 21 | Studying Acid Precipitation

One investigation that you can carry out will answer the question of whether the precipitation in your area is acidic or not. The materials you will need are simple and easy to get.

Materials
A plastic container
pH paper
Notebook
Pen
Bottle of deionized water

You will need a plastic container in which to collect the precipitation. Also, you will need pH paper (litmus paper) which changes colors between pH 1 and pH 6 or between pH 2 and pH 7. Either will work nicely. (A color key is provided with the strips of pH paper so you can determine the pH of rain or snow. After the paper becomes wet with the precipitation, you compare the color change with the colors on the key.) A small supply of pH paper can be purchased from a chemical supply company or possibly a pharmacy in your neighborhood. You will also need a notebook and a pen to record your observations. The last item you will need is a bottle of deionized water for rinsing out your collecting container after each use.

Pure rain or snow is slightly acidic (pH 5.6). In the western portion of the United States, the average pH of the rain is about 5.5. In the eastern United States, the rain is more acidic, approximately 4.5. Remember, the lower the pH, the more acid the precipitation.

You should take observations of precipitation events over several months or longer, and your data should include:

- Date/time of day
- Direction of the wind
- Duration of the precipitation event
- Type of precipitation (rain, snow, etc.)
- pH of the precipitation

Human Ecology

Questions to Consider

1. From which direction does the precipitation with the lowest pH come? How can you explain this observation? _____
   _____
   _____

2. During which season is acid precipitation the worst? What is a possible explanation for this? _____
   _____
   _____

3. How does the pH of local streams, ponds, lakes, and rivers compare with the average pH of the precipitation in your area? _____
   _____

4. Is any type of precipitation more acidic than the others? _____ If so, which one?
   _____
   _____

5. What is the average pH of the precipitation in your area during the time you kept records? _____

Human Ecology

CHAPTER 8 REVIEW | Human Ecology

1. Name the major air pollutants and their sources. _____
   _____
   _____
   _____

2. How do regional haze and smog differ? _____
   _____
   _____

3. Why doesn't the western United States have a problem with acid deposition?
   _____
   _____
   _____

4. Why are power plants and oil companies reluctant to remove pollutants from their fuels? _____
   _____
   _____

5. Why will the carrying capacity of a desert be less than that of a temperate forest?
   _____
   _____
   _____

GLOBAL WARMING | 9

There is no doubt that our earth's atmosphere is warming up. The hottest year on record was 1998. The next year, 1999, was cooler, but it still was the sixth hottest year. The graph below demonstrates that global warming is really happening.

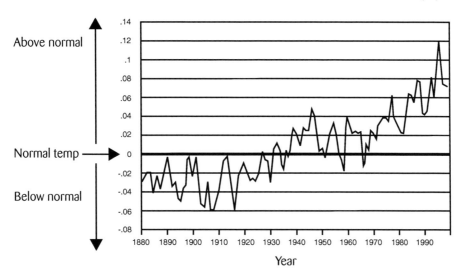

Let's take a closer look at this graph in order to understand what it tells us. First, you may be wondering why the title uses the word *deviation* rather than *actual global temperatures*. This is because an absolute global temperature is very hard to determine. The absolute temperature would change quickly as you climbed up a mountain or down into a valley. It is easier to use the change (temperature deviation) from year to year. In the years between 1951 and 1980, the global mean temperature is 13.9 °C. That figure is still being used today.

What is that temperature in degrees Fahrenheit? _____

Global Warming

The global mean temperature has gone up 0.6 °C since the 1880s. That may not seem like much, but the graph shows how the temperature has deviated *above* the mean in recent decades. Now let's look a little more closely at the graph.

In what way is the *X* axis unusual? _____
_____
_____

What was the temperature deviation in the year 1900? _____ What does this mean?
_____
_____
_____

Approximately when did the warming trend begin? _____
What were the three coldest years in the twentieth century? _____
_____

Compare the temperature deviation of the 1950s to that of the 1980s.
_____
_____
_____
_____
_____

You may wonder why there should be such a fuss about the average surface temperature going up less than one degree Celsius. What can such a small change do? Let's see what has happened thus far because of the temperature change, and what may happen if this trend is not stopped or reversed.

THE EFFECTS OF GLOBAL WARMING

The earth's northern regions have already become greener by more than 10 percent since 1980. Not only are more square kilometers greener, but the plants stay greener longer. Spring comes at least a week earlier and fall arrives later, so the growing season is about 15 days longer. This doesn't sound bad. More food can be grown. So far, so good for the northern latitudes. However, in the southern regions of the United States and in those areas close to the equator, the soil will dry out more quickly due to the heat. Therefore, food production will decrease in these regions.

The ice cap of Mount Kilimanjaro in Tanzania has lost 82 percent of its ice since it was first measured in 1912. Mountain glaciers have been melting all over the world. The same is true of the polar ice cap at the North Pole. Polar bears are in danger of starvation as melting ice eliminates solid surfaces to walk on and keeps them from their food.

Polar bears on ice. (Corel CD)

The result will be higher sea levels all over the world. The sea level has risen somewhere between 10 and 20 cm over the past century, and scientists predict a rise of up to 100 cm in the current century. This projected rise of one meter will cause coastal flooding and beach erosion all over the world. Many countries will have to build sea walls for protection or force people to evacuate beach areas. In addition, German scientists recently reported that the seas are getting rougher with higher waves. They have matched the increase in wave height to increased air temperatures near the earth's surface.

Global Warming

Melt pool and iceberg. (Corel CD)

Global warming will increase the rate of evaporation from the land, causing drier soils and more deserts. Water will be exchanged faster between the oceans, the atmosphere, and the land, resulting in more frequent floods and droughts. This trend may already have started; in the early 1990s there were two severe floods in the Midwest in a five-year period.

Health is an important concern in relation to global warming. First, severe heat waves will be more frequent. In 1995, Chicago recorded 495 deaths due to heat stress. As the earth grows warmer, what will happen to the number of deaths from heat stress? Warmer weather equals fewer "killing frosts," so there will also be a spread of mosquitoes and other insects.

Name three diseases that are transmitted by insect bites. _____
_____
_____

Since warmer air will produce more mold spores and pollen in the atmosphere, cases of asthma and other plant allergies will also be on the increase.

Forest and grassland changes due to global warming will affect the organisms in most ecosystems. In the following table, describe what changes might occur in each of the listed ecosystems.

Ecosystem	Change
Tundra	
Desert	
Mountain regions	
Wetlands	

CAUSES OF GLOBAL WARMING

Currently, there are three theories about why global warming is taking place. However, most scientists believe that the cause is an increase in the **"greenhouse gases."**

Carbon dioxide (CO_2) is a major greenhouse gas in the atmosphere. In fact, without any carbon dioxide in the atmosphere, the earth would be a much colder place to live. The global mean temperature would be below 0 °C instead of being close to a comfortable 14 °C. Most carbon dioxide comes from the decomposition of dead plants and animals and the respiration of living animals and plants. For thousands of years there was no problem with this, because the oceans absorbed much of this CO_2, taking it out of the atmosphere. Plants carrying on photosynthesis also absorbed a great deal of the atmospheric carbon dioxide.

Everything was fine until people began to release huge quantities of additional carbon dioxide into the atmosphere, beginning in the late nineteenth century and increasing dramatically up through the early twenty-first century. Auto engines, power plants, industrial mills, and home and business heating systems burn coal, oil, or natural gas. This accounts for 98 percent of the CO_2 added to the atmosphere. The other 2 percent is due to increased deforestation and mining. To understand how carbon dioxide builds up in the atmosphere, we can take another look at the carbon cycle.

Global Warming

During the period 1981–1990, 7.1 billion metric tons (BMT) of carbon dioxide was released worldwide. Of that amount, 5.5 BMTs were the result of the burning of fossil fuels. An additional 1.6 BMTs came from the destruction of tropical forests and other forms of vegetation, respiration, and decomposition.

1.8 BMTs were removed from the atmosphere by existing plants (land and aquatic) through photosynthesis. Another 2.0 BMTs were absorbed by oceans, lakes, etc. Both of these sources of CO_2 removal are known as **sinks**. However, *3.3 BMTs of carbon dioxide remained in the atmosphere.*

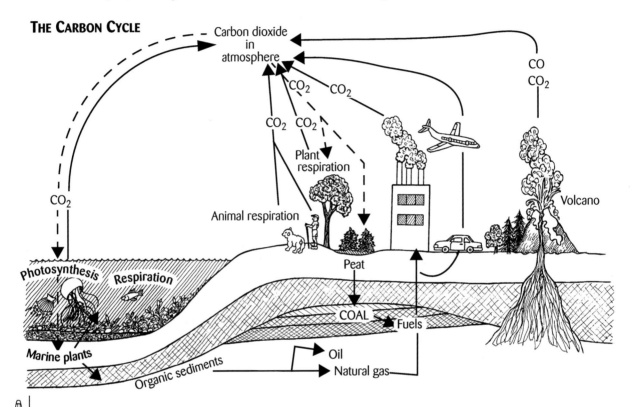

THE CARBON CYCLE

Be sure to read the text in the upper left corner of the diagram. It explains how carbon dioxide is added to the atmosphere. Then read all the text in the diagram, and follow the arrows. When you have finished, answer the following questions.

1. Name three sources of carbon dioxide _____

2. Which ones can we control? _____

3. What do carbon dioxide sinks do? _____

4. Explain why carbon dioxide tends to build up in the atmosphere. _____

© 1979, 1989, 2001 J. Weston Walch, Publisher

Global Warming

Methane (CH_4) is another of the major greenhouse gases. Methane molecules trap 20 times more heat than carbon dioxide molecules. Methane is released from wetlands, decomposing wastes in landfills, rice paddies, the stomachs of cows and sheep, and termites. It also leaks out of coal mines and natural-gas deposits. There are about two *billion* cows and sheep on the earth today, each belching up 200 grams of methane or more each day!

In the last few years, the amount of methane being added to the atmosphere has been decreasing. Scientists are not positive why this is so. One theory is that global warming is causing the wetlands—the number one methane source—to dry up, resulting in a reduction of methane production. Another possibility is that since the outbreak of "mad cow" disease and hoof-and-mouth disease in Europe, less beef has been sold; thus, fewer cows have been raised. In the United States, beef consumption is also down, but for a different reason. Many Americans have cut down on their consumption of hamburgers and other meat products that are high in fat content in an effort to lose weight or cut their cholesterol levels.

Next in order of importance of the greenhouse gases is nitrous oxide (N_2O). This gas is added to the atmosphere when bacteria break down human and animal waste products. Nitrogen fertilizers breaking down in the soil also release nitrous oxide into the atmosphere. The burning of oil, coal, and natural gas also creates nitrous oxide. Since the mid-1700s, nitrous oxide content in the earth's atmosphere has increased by more than 7 percent. A molecule of nitrous oxide absorbs 270 times more heat than a carbon dioxide molecule!

Some gases of minor importance, such as sulfur hexafluoride, also contribute to global warming. These gases are not included in the table below, which shows greenhouse gas concentrations in the atmosphere at the beginning and the end of the twentieth century.

Major Greenhouse Gas	Concentration in the Atmosphere in 1900	Concentration in the Atmosphere in 2000
Carbon dioxide	300 parts/million	375 parts/million
*Methane	900 parts/billion	1,800 parts/billion
Nitrous oxide	275 parts/billion	325 parts/billion

*Since 1980, the rate of increase of methane in the atmosphere has been negative. That is, each year less methane is added.

Which greenhouse gas is most prevalent in the atmosphere today?
_____

Using the data in the table, explain why carbon dioxide is the most important greenhouse gas.
_____
_____
_____

Global Warming

THE GREENHOUSE EFFECT

For millions of years, greenhouse gases have been at work regulating the heat entering our atmosphere. Scientists call this the **greenhouse effect.** The greenhouse effect is not a new phenomenon. Remember, without the greenhouse effect, our earth's average surface temperature would be freezing.

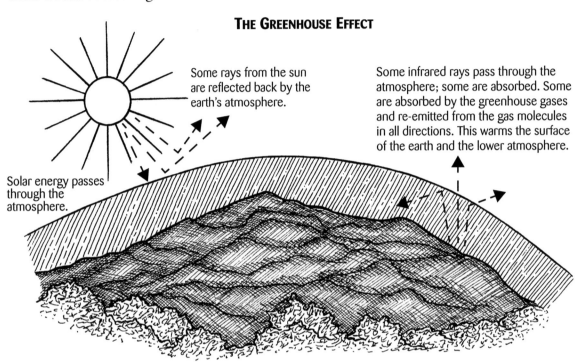

Here is how the greenhouse effect works. Energy from the sun, mainly in the form of visible light, arrives at the earth's atmosphere. Only two thirds of this light is absorbed by the earth's surface and atmosphere. The other one third is immediately reflected by clouds back into space. The two thirds that were absorbed by the earth's surface is then re-emitted back into the earth's atmosphere—not as light, but as infrared radiation (heat).

The greenhouse gases are completely transparent to light rays, but they are *not* completely transparent to heat and can absorb some of it. In other words, they trap the heat in the earth's atmosphere, but they don't trap the light. If you have ever sat in a closed automobile that is parked in the sun on a cold day, you have experienced the greenhouse effect. The glass windows allow light to enter, and the sunlight is absorbed by you, the seats, and so on. That energy is re-emitted as heat. The glass prevents the heat from escaping, and so the car warms up.

A second theory is that black carbon soot **aerosol** is responsible for global warming. A black soot aerosol is composed of microscopic particles in smoke or fog. Actually, black soot is not a gas. It is produced by the burning of fossil fuels and forest fires. When some trucks or buses start their engines, a cloud of exhaust gas is released. That cloud is a black soot aerosol. About 10 million tons of black soot enter the atmosphere each year. Ozone, which has been mentioned before, is also a part of black soot aerosols. It is formed when sunlight strikes the smoke produced by burning fossil fuels or forest fires. Ozone is a poisonous compound; you may recall weather reports mentioning ozone alerts when conditions are good for its formation. Bear in mind that the ozone in the ozone layer which is found high up in the stratosphere was formed there; it did not originate on the earth's surface. No matter where it is found in the atmosphere, near the surface or high up in the stratosphere, ozone traps heat.

The third theory states that we are not responsible for global warming, the sun is! The scientists who support this theory state that the sun has storms and seasons, with an eleven-year cycle of sunspots and solar flares. In addition to visible light, the sun emits ultraviolet light, which can split oxygen molecules into oxygen atoms. These then combine with other oxygen molecules to form molecules of ozone. Furthermore, solar wind and magnetic fields of the sun, which cause the Northern Lights, also change the flow of cosmic rays from outer space into our atmosphere. This in turn affects cloud formation. Clouds also trap heat in the atmosphere. You may have realized that cloudy nights tend to be warmer than clear nights at the same time of the year.

As of this writing, the strongest case made with scientific data supports the greenhouse gases theory.

What Can We Do to Cut the Emission of Greenhouse Gases?

Here are some ideas:

1. Use solar power, wind power, and hydroelectric power when possible.
2. Raise the pollution standards for trucks and SUVs to that of automobiles.
3. Tighten the limits of pollution for oil-fired and coal-fired power plants.
4. Require hybrid motors and hydrogen fuel cells be used to power cars. (These topics will be considered in another section of this book.)
5. Plant forests.
6. Convert crop land into grassland.
7. Switch from high-carbon coal to low-carbon coal.

Now it is your turn to write some possible solutions.

_____
_____
_____
_____
_____

Global Warming

ACTIVITY 22 | Demonstrating the Greenhouse Effect

In this activity, you will create two models of the earth's atmosphere. Then you will determine how light is transformed into heat in each model.

> **Materials**
> Two 1-liter soda bottles made of clear plastic with paper labels removed
> Gas burner
> Phillips head screwdriver
> Safety goggles
> Two metric thermometers
> Transparent tape

Procedure

Note: The first two steps must be done under adult supervision while wearing goggles.

1. Lay one of the soda bottles on the tabletop. Heat the tip of the screwdriver using the gas burner. The screwdriver need not get red hot. While holding the soda bottle by its neck, use the heated screwdriver tip to melt seven holes in a line down one side of the bottle as shown in Diagram A.

    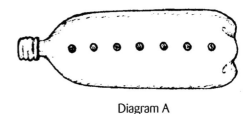
    Diagram A

2. Repeat the first step, only this time, prepare 13 holes as shown in Diagram B.

3. Set the plastic bottles on their sides so that the holes are facing up. Indoors, place them on a tabletop that is in the sun. If you are doing this outdoors, be sure that there is little or no wind. Insert a thermometer in the

    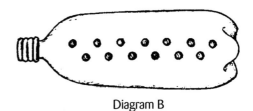
    Diagram B

    middle hole in each bottle. Use two thin books to keep the bottles from rolling around. Use the transparent tape to hold the thermometers in place. Be sure that the thermometer tips are near the middle of the inside of the bottles and are not touching the plastic on the side of the bottle that is resting on the flat surface.

4. Record your initial temperatures. Record the data in the table on the next page.

Global Warming

5. Continue to take readings for 30 minutes, recording your data.

Reading	7-Hole Bottle	13-Hole Bottle
Initial	°C	°C
5 minutes	°C	°C
10 minutes	°C	°C
15 minutes	°C	°C
20 minutes	°C	°C
25 minutes	°C	°C
30 minutes	°C	°C

6. Graph your data in a line graph. Use different colored ink for each bottle's temperature.

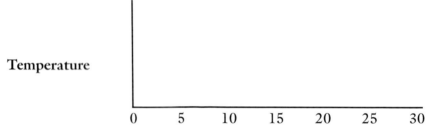

Analysis

1. What conclusions can you reach from comparing the data for each bottle?
   _____
   _____

2. How do you explain the differences in the data for each bottle? _____
   _____
   _____

3. The bottles are models of the atmosphere. (The plastic acts like glass when it comes to light and heat.) Which part of the model represents the greenhouse gases?
   _____
   _____

4. Which model represents an atmosphere with the most greenhouse gases in it?
   _____
   _____

5. Why would your results *not* be valid if you did this outside on a windy day?
   _____
   _____

CHAPTER 9 REVIEW | Global Warming

1. Explain why global warming is a serious problem to you and to future generations.

   _____
   _____
   _____

2. Why might replacing all gasoline engines with hydrogen fuel cells reduce global warming? _____
   _____
   _____

3. What is the greenhouse effect? _____
   _____
   _____

4. State and explain three ways by which carbon dioxide emissions could be reduced.

   _____
   _____
   _____

5. What are the sources and sinks of carbon dioxide? _____
   _____
   _____

CAR TALK | 10

IS BIGGER BETTER?

In the early twenty-first century, citizens in most industrialized countries should feel deeply concerned about what their industries are doing without regard for the ecosystem—or the welfare of future generations. It will take enormous effort, and government enforcement, to begin reversing the tide of global warming and resource depletion.

Let's see how one auto manufacturing firm affects the ecosystem now. This company plans to sell a giant passenger vehicle in a couple of years. The company believes that bigger is better. This vehicle will be taller than a U.S. Army tank. It will be 20 feet long—almost twice the size of a compact car. It will weigh the same as a Tyrannosaurus rex weighed, about 12,600 pounds. This vehicle will only get 10 miles per gallon of fuel. Due to its weight, it will be classified as a medium-duty truck by the government.

Trucks don't have to meet the same standards for air pollution controls as automobiles. However, some states will ban this monster because its great weight would be damaging to bridges and roads. It will also lack an effective **catalytic converter**, a device that changes nitrogen oxides and carbon monoxide into less harmful exhaust gases. (Both nitrogen oxides and carbon monoxide cause smog.) A catalytic converter adds to the cost of any vehicle. This is because palladium (a precious metal), used to make catalytic converters, now costs about $35 per gram. For a small car, the catalytic converter adds $500 to its price. The bigger the vehicle, the larger and more expensive the catalytic converter has to be.

In addition, due to its great weight, this vehicle won't have to meet the strict safety standards or fuel economy (miles per gallon) standards that cars do. Fuel economy is important when you consider that the burning of fuel releases smog-causing gases as well as hydrocarbons, contributing to global warming. Consider the following: If a sports utility vehicle (SUV) gets 15 miles to the gallon, how many gallons would it need to travel 750 miles? A small car gets 30 miles per gallon. It will use half as many gallons of gasoline to make the same trip.

Because of pressure from the auto industry, the government has actually *eased* pollution controls and fuel economy regulations on sports utility vehicles. They are classified as light trucks, which pollute the air more than cars. The trucking industry is not interested in preserving the atmosphere, because applying car standards to trucks would mean smaller profits for the truck manufacturers and related industries.

Car Talk

Examine the chart below.

Vehicle Type (Year 2000–2003)	Grams of Smog-Causing Emissions per Mile
Mid-size passenger cars	3.7
Light trucks and SUVs	5.2
Heavy truck	6.5

Now convert the data into a bar graph using the space below.

Grams of smog-causing emissions

Vehicle type

Keep this information in mind: Motor vehicles release more than 50 percent of the smog-causing gases and hazardous air pollutants on the planet. Think back to the sports utility vehicle making the 750-mile trip. How many grams of smog-causing pollutants would it release during that trip? (Use the table above.) _____

This year, motor vehicles are expected to be driven 4,000,000,000 miles throughout the United States. Using the table above, imagine that all of the SUVs in the U.S. were driven 1 trillion miles, all mid-size passenger vehicles were driven 2 trillion miles, and heavy trucks were driven 1 trillion miles.

How many kilograms of smog-causing emissions would be released by American vehicles into the atmosphere this year? _____

What would you think of a car that got 60 miles per gallon of gasoline, or one that released *zero* pollutants into the air? Such cars will be discussed later in this chapter. The trend in all industrialized countries should be to burn less fuel so that the amount of carbon dioxide and other harmful gases escaping into the atmosphere will be reduced. Apparently, the makers of SUVs and even bigger vehicles are failing to meet this responsibility.

HYBRID AUTOMOBILES

Some auto makers *are* making efforts to address the situation. A handful of Japanese and American car makers are manufacturing **"hybrid" cars.** A hybrid runs on both gasoline and electricity; the use of electricity greatly reduces gasoline consumption. Hybrid cars get between 50 and 65 miles per gallon. There are two kinds of hybrid. One kind is called a "mild hybrid" or "mini-hybrid." The second type is known as a "full hybrid."

The mild hybrid has a 6-cylinder engine and a 12-volt battery for running the lights and wipers. (This is the same as is found in a typical car.) It also has a 42-volt battery pack. This vehicle is powered by both electricity from the battery pack and by the gasoline engine. When the car comes to a stop, the engine shuts off automatically. When the driver steps on the accelerator, the engine restarts. When the driver applies the brakes, recharging of the batteries occurs.

The full hybrid has the usual 12-volt battery, a small 4-cylinder engine, and a 300-volt battery pack. Its engine also shuts off automatically whenever the car comes to a stop. The engine restarts when the driver steps on the gas pedal. However, at low speeds the full hybrid is powered only from the battery. When the driver applies the brakes, the batteries are recharged.

Compare the mild hybrid and the full hybrid in as many ways as you can. How are they alike? How do they differ?

Similarities	Differences

Both the mild hybrid and the full hybrid cars run on battery power and gasoline. When operating on gasoline, they—like all gasoline-powered vehicles—are only about 25 percent efficient. This means that 25 percent of the energy in the gasoline is used to move the vehicle. The other 75 percent is changed into heat or lost to friction. This explains why cooling systems are so important in cars with internal combustion engines.

There are some cars on the road that are powered only by batteries. They emit *zero* gases because they aren't powered by gasoline. Unfortunately, battery-powered cars have a few problems. The batteries are quite large and their traveling range is short, requiring frequent recharging. It takes a long time to recharge these batteries.

FUEL CELL ENERGY

A completely different kind of energy is produced by **fuel cells.** Fuel cells produce electricity through a chemical reaction, similar to the way in which batteries produce electricity. However, the fuel cell, unlike the battery, only produces electricity when it receives fuel. There is no burning, and the chemical production of electricity takes place at low temperatures. Different fuels can be used, but the best fuel is hydrogen. Methanol and natural gas could also be used, but these contribute to air pollution. The fuel is stored in a separate tank like the gas tank of a car. In fuel cells, hydrogen and air react, producing electricity, water, and some heat.

Some of the advantages of fuel cells are:

- no moving parts
- clean
- quiet
- produce less heat (85 °C compared with 125 °C for a car engine)
- can use different fuels
- greater efficiency

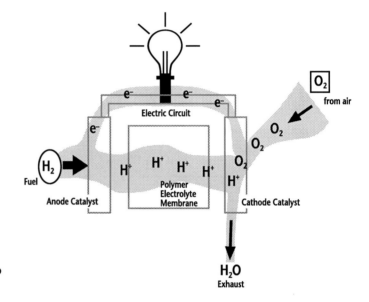

The main part of a fuel cell consists of a platinum-coated membrane electrode assembly, with a positive electrode bonded to one side and a negative electrode to the other. Hydrogen gas flows to the positive electrode through special openings in flow-field plates. The platinum acts as a **catalyst** to separate the hydrogen atoms into protons and electrons. The electrons move through an outside circuit and end up on the other side at the negative electrode. The movement of electrons (electric current) through the outside circuit provides the power source. The protons pass through the membrane to the negative electrode. Air flows through other channels in the flow-field plates to the negative electrode. At the negative electrode, oxygen in the air, the protons, and the electrons from the outside circuit combine to form water vapor and heat.

Fuel cells are arranged in stacks. The more cells in the stack, the higher the voltage. The larger the cells, the greater the current. Fuel cells can be small enough to power a cellular phone or large enough to provide a small city or town with electricity. Fuel cells can be used for many purposes, including producing electricity and water for the astronauts during space shuttle missions. In 1999, an experimental car powered by fuel cells was produced by one of the automobile manufacturers. It was able to travel 250 miles without refueling. Its top speed was 90 miles per hour.

ANOTHER SOURCE OF ENERGY

Perhaps you would like this automobile? It has a top speed of 78 miles per hour and can move along at 55 miles per hour using the same amount of electricity as a household hair dryer. It has no headlights, though, so you can only drive it in the daylight. How do you think this car is powered? _____

_____

If you thought that it was powered by sunlight, you are correct. This car was designed and built by students at the University of Minnesota. Experimental solar cars have been built by school students as well as by people in large companies.

Electric motors power **solar** cars. These motors are much more efficient than gasoline engines, which, as was mentioned earlier, waste 75 percent of their energy to heat and friction. Solar cars are powered by solar (photovoltaic) cells. You probably know what *photo* means ("light"). You should also be able to figure out the meaning of *voltaic* ("electricity").

Photovoltaic (PV) cells are made of **semiconductors.** These are made with the elements silicon, phosphorus, and boron, which are fashioned into a thin sheet. Then, the semiconductor sheet is covered with clear glass. A back plate and metal electrodes complete the PV cell. When light energy enters the cell, some of that energy (about 15–25 percent) is absorbed into the semiconductor. Electrons in the semiconductor are knocked loose by the incoming light energy. If metal contacts are placed on the bottom of the PV cell, the flowing electrons can be used to power a calculator. Link up many PV cells and you can power a car.

The construction of a PV cell is more complicated than this description implies, but you have the basic idea. Unfortunately, PV cells are not too efficient in terms of producing electricity. They can absorb and use only about 15 percent of the sun's energy. This is because sunlight is composed of many different wavelengths of light that form the visible spectrum. Some wavelengths pass right through without being absorbed, while others have so much energy that some of this energy is not absorbed but changed to heat.

Car Talk

ACTIVITY 23 | Testing a Solar-Powered Model Car

1. Purchase a kit containing all the materials you need to make a working solar cell, starting with a sheet of silicon. Use it to make a model solar car. Or, you can purchase a solar car kit from a scientific supply company. Your school may have one.

2. Take the car outdoors on a sunny day. The car has a pivot on which the PV cell is mounted.

3. Determine which angle to the sun propels the car the fastest. To do this, set a distance for the car to travel; 3 meters will work. Have a classmate time the car's trips using a stopwatch. It would be wise to make three runs for each angle and then use the average times. Fill in the data table below.

Angle to Sun	Sunlight	Artificial Light
180° (horizontal)		
150°		
120°		
90° (vertical)		

4. What relationship exists between the angle of the PV cell and the sun's position in the sky? _____

5. Try adding a lens to focus the sunlight on the cell. What was the effect upon the car's speed? _____

   What problems did you have keeping the sunlight focused on the PV cell?
   _____

6. Take the car indoors. Using a halogen lamp as the light source, repeat Step 3. How does the car's speed under artificial light compare with its speed under natural sunlight?
   _____

7. Use colored cellophane to cover the halogen lamp. Determine which color powers the car the fastest. Try red, yellow, blue, and green; add any other colors of your choice. Arrange your data in a table like the one below.

Colors	Average Time to Travel 3 Meters
Red	
Yellow	
Blue	
Green	

8. **Optional:** With your teacher's permission, try to improve on the design of the solar car. If you are successful, you can visit web sites that sponsor competitions for solar cars built by high-school students.

Car Talk

ACTIVITY 24 | Pollution in the Business District

In this activity, you will calculate traffic flow at a busy intersection in your community's business district at varying times of day. Record the data in *numbers of vehicles per hour*. Use the chart that follows as a model.

Location of intersection: _____

	No. of CARS/hr	No. of BUSES/hr	No. of TRUCKS/hr	No. of VANS/hr	No. of TAXIS/hr
MORNING Time:					
AFTERNOON Time:					
EVENING Time:					
NIGHT Time:					
AVERAGE					

1. Do you think the average data you collected are typical for your community? Explain your answer. _____

_____

_____

2. What factors in the community influence the volume of traffic that exists? (Consider highway patterns, population density, factories, shopping centers, businesses.)

_____

_____

_____

3. What evidence of pollution due to traffic did you find? (Include odors, smoke, noise, increase in temperature, and dead or dying plants near the intersection.)

_____

_____

_____

© 1979, 1989, 2001 J. Weston Walch, Publisher   *Understanding Basic Ecological Concepts*

4. What laws do you feel should be passed to reduce pollution due to automotive traffic?

_____
_____
_____

5. What devices for controlling air and noise pollution are used in your community?

_____
_____
_____

6. How do parking meters, one-way streets, pedestrian safety lanes, shrubs, and trees in the business district help reduce pollution and solve environmental problems?

_____
_____
_____

7. Describe the sanitary conditions of the business district. Are there rats, roaches, or littering problems? If so, what can be done about them? _____

_____
_____

8. What environmental pollution problems are experienced by merchants, shoppers, and motorists? Be sure to interview people in each category. _____

_____
_____
_____

9. After gathering all of your data and answering the questions, what suggestions can you and your class make to help solve some of the problems you have observed?

_____
_____
_____

Car Talk

Going One Step Further

Interview the chief of police, head of the traffic department, mayor, and/or members of the chamber of commerce to discover what plans have been proposed in your community to eliminate present and future environmental problems in the business district. Record the notes of your interview on the following lines.

_____
_____
_____
_____
_____
_____
_____
_____
_____
_____

Like other industrialized countries, the United States uses mainly coal, oil, and natural gas for its energy. These **fossil fuels** are expensive, as the people in California found out in 2001. Besides being expensive, these fossil fuels also seriously damage the environment. In addition, they will become scarce, since the planet will eventually run out of such resources.

Our government should be concentrating on developing other sources of energy. Solar energy will be around forever, and hydrogen for fuel cells will also be available for our needs for a very long time. In addition, there are other sources of energy that could be tapped, including the wind and ocean waves. Tidal energy is available on a continuous basis. It is time to pursue these avenues more aggressively and begin decreasing our dependence on dangerous fossil fuels.

Chapter 10 Review | Global Warming

1. Why do the manufacturers of SUVs want them classified as light trucks?

   _____

   _____

2. What is the advantage of having a vehicle's engine shut off when waiting for a red light to change to green? _____

   _____

3. Why are hybrid cars a forward step in reducing smog and global warming?

   _____

   _____

4. Why aren't battery-powered vehicles being manufactured on a large scale?

   _____

   _____

5. Which is better: a car powered by fuel cells or one powered by solar energy? Explain your reasoning. _____

   _____

   _____

STRATOSPHERIC OZONE LOSS — 11

The word *stratospheric* in the title of this chapter refers to the **stratosphere,** which is the uppermost layer of Earth's atmosphere. The stratosphere starts about 10 kilometers above the earth's surface and continues to approximately 50 km high. In the region between 30 and 40 km above the earth is where stratospheric ozone forms the **ozone layer.** This region is above the part of the stratosphere in which jet planes fly.

Ozone is a gas very similar chemically to the oxygen we breathe. That oxygen element contains two atoms of oxygen; ozone contains three. In terms of its physical and chemical properties, though, ozone molecules are far different from oxygen molecules. Ozone is a blue poisonous gas that has a noticeable odor. You probably have smelled ozone if you have ever been near a sparking electric motor. (In fact, ozone is prepared in the laboratory in a similar manner.)

Ozone is a Dr. Jekyll and Mr. Hyde molecule. Close to the earth's surface, large concentrations of ozone are deadly to living things. It helps form smog and contributes to the greenhouse effect. Fortunately, only 10 percent of the ozone in our atmosphere is found close to the earth's surface (as discussed in the Global Warming chapter). The remainder is found in the upper altitudes of the stratosphere, where ozone does a good job of absorbing harmful **ultraviolet (UV) radiation** released from the sun. This radiation has been linked to several types of skin cancer and cataracts in humans. A cataract changes the lens of an eye from clear to hazy, making it difficult to see.

Ultraviolet rays kill marine plants. Remember that green plants are the foundation of all food webs. If they die, so do the consumers! The soybean is one example of a land plant whose yield decreases as ultraviolet radiation increases. Soybeans are an important crop in many countries, including the United States.

Compare oxygen with ozone by completing the table below:

Oxygen	Ozone
	A blue gas
Odorless	
Composed of two atoms of oxygen	
	In the stratosphere, absorbs dangerous solar ultraviolet radiation
	Poisonous to living plants and animals close to Earth's surface

Stratospheric Ozone Loss

THE OZONE CYCLE

It is important to recall that stratospheric ozone does not originate on the earth's surface. Read the text and look at the diagram that follows to understand the ozone cycle. The numbers in parentheses indicate the portion of the diagram that illustrates that part of the text.

(1) Ozone molecules in the ozone layer are formed when ultraviolet (UV) rays from the sun collide with oxygen molecules in the stratosphere. This causes each oxygen molecule to split into two free oxygen atoms. (2) Almost immediately, these two oxygen atoms will bond with two nearby oxygen molecules, forming two molecules of ozone. (3) The ozone molecules absorb UV rays from the sun. Sometimes the UV radiation will split each of these ozone molecules into an oxygen atom and an oxygen molecule, while the rest of the energy is changed to heat. (4) The oxygen atoms will then recombine with oxygen molecules, reforming ozone molecules, and the cycle continues.

Based on the explanation above, it would seem that the molecules of ozone would continually increase because of the sun's action upon oxygen molecules in the stratosphere. However, this is not the case. In fact, stratospheric ozone molecules are constantly being destroyed by reactions with other chemicals that have been present in the stratosphere since before humans appeared on Earth. Some of these ozone-destroying chemicals are nitrogen from the soil and oceans, hydrogen from water vapor, and chlorine from the oceans. Thus, the number of ozone molecules stayed fairly constant for millions of years.

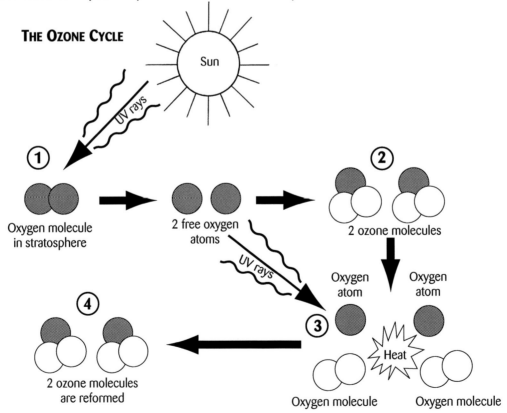

OZONE LOSS

Sadly, this balance changed around 1930. At that time, a class of chemicals known as **chlorofluorocarbons (CFCs)** were synthesized (*chloro* = chlorine atoms, *fluoro* = fluorine atoms, *carbons* = carbon atoms). CFCs had many unusual and wonderful properties. They were not poisonous and didn't corrode. They weren't flammable and did not react with other substances. Beginning in about 1935, they were widely used as coolants, replacing ammonia in refrigerators. Later, CFCs were used in home and auto air conditioners. In the 1950s and 1960s, CFCs were used as propellants in spray-paint cans, hair sprays, and other spray products. They also trapped heat very well and were used to make foam coffee cups. By 1970, about 1 million tons of CFCs were being used each year for all of these purposes, as well as in the electronics industry, where they are used as cleansers.

In the lower layers of the atmosphere, up to 29 km above the earth, CFCs present no problem. Scientists believed that they were harmless, because close to Earth's surface they were practically insoluble in water and were not affected by sunlight. However, when CFCs drift up to the stratosphere, 30 km or higher, they are broken apart by the sun's energy. That has resulted in extra chlorine atoms being added to the stratosphere.

The diagram below illustrates how chlorine atoms affect ozone molecules. In essence, chlorine destroys ozone by means of a complicated chain reaction. Here is a startling fact: *One atom of chlorine* can eventually destroy *100,000 ozone molecules*. To make matters worse, CFC molecules can remain in the stratosphere for up to 100 years. Bear in mind that as the number of ozone molecules in the stratosphere decreases, people, animals, and plants are placed at increased risk of disease and even death.

THE EFFECT OF CHLORINE ATOMS ON OZONE

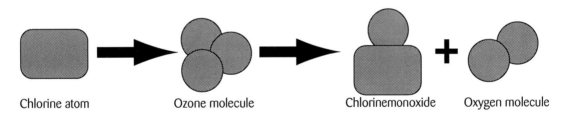

Chlorine atom → Ozone molecule → Chlorinemonoxide + Oxygen molecule

One chlorine atom combines with one oxygen atom of an ozone molecule, forming chlorinemonoxide. This causes an oxygen molecule to be released and the ozone to be destroyed.

Stratospheric Ozone Loss

You have probably heard about the hole in the ozone layer. This term is somewhat misleading. In fact, most stratospheric ozone is found in a band between 30 km and 40 km above the earth. The "hole" is not really a hole—like the hole you see in the center of a doughnut or a CD. Rather, it is a *depression*. It is an area of the stratosphere that is about 15 million square kilometers (the size of the North American continent) with very little ozone in it. Look at Graph 1 below. Notice that the "hole" doesn't go all the way through to the bottom of the ozone layer.

Ozone loss was first discovered in the stratosphere over the Antarctic in 1970. Over Antarctica, the loss is greater than over other parts of the earth. Thus, it is easier for scientists to study ozone loss over the Antarctic region than over other regions. Scientists have noted that in some years, ozone levels over the Antarctic have fallen by more than 60 percent. Over the United States, depending on the season, ozone levels have dwindled by as much as 10 percent. All over the world, ozone levels have declined, making this a global problem rather than a national or local problem. Remember that the less ozone in the stratosphere, the less protection we have from deadly ultraviolet radiation. Indeed, during the last half of the twentieth century, the world saw the rate of **malignant** (deadly) skin cancer increase 10 times over the previous 50 years. Much of that increase can be blamed on the loss of ozone.

Look at Graph 1, which shows the ozone layer over Antarctica for part of the year 2000.

For how many months does the "hole" last?_____

During which month is the ozone loss the greatest?_____

Where are the ozone molecules found, above or below the line in the graph?
_____

Explain.

GRAPH 1: DEPTH OF THE OZONE LAYER

Stratospheric Ozone Loss

GRAPH 2: RELATIVE AMOUNT OF STRATOSPHERIC OZONE OVER ANTARCTICA DURING THE MONTH OF OCTOBER

GRAPH 3: RELATIVE INCREASE/DECREASE OF CFCs IN THE STRATOSPHERE

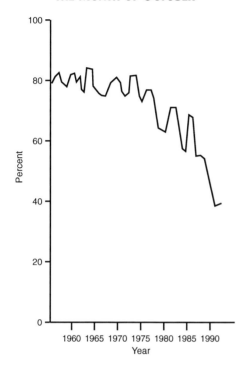
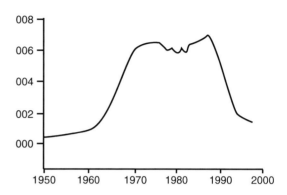

Now look at Graphs 2 and 3.

Look at the years from 1960 to 1970 on both graphs. How do they compare? _____

_____

_____

What possible conclusion can you reach from looking at these two graphs? _____

_____

_____

By the middle of the 1970s, scientific experts were convinced that the CFCs were a threat to the world. In 1979, the United States and other countries banned the sale of aerosol cans (spray cans). Unfortunately, the ban did not include CFCs used for coolants in air conditioners and other machines. To make matters worse, many countries continued to release CFCs into the atmosphere from spray cans.

Stratospheric Ozone Loss

Examine Graph 2 again. What effect did this release of CFCs have on the amount of ozone over the Antarctic in October of the early 1980s? _____
_____

The worldwide use of CFCs was increasing at a rate of 3 percent per year starting in 1985. What evidence of that increase do you see in Graph 2? _____
_____

In Graph 3? _____
_____

In 1987, an agreement was signed in Montreal, Canada, which cut the use of CFCs on a worldwide basis. Furthermore, the nations of the world agreed to completely ban the production of CFCs by the year 1996.
How is this reflected in Graph 3? _____
_____

What was the effect on the ozone over Antarctica as shown in Graph 2? How can you explain this from what you have read so far? _____
_____
_____

   At the close of the twentieth century, hopes were high for the ozone layer in the new century. However, those hopes were dashed in 1999, when it was discovered that illegal CFCs were—and still are—being made in Russia, China, and India. A staggering 15 million tons have been smuggled around the world. In 1999, United States customs agents seized 1,000 metric tons of CFCs and convicted 17 people for smuggling these banned chemicals into the country. As these chemicals become scarcer and scarcer, the "black market" price goes up—and there are greedy people who care more for money than for their own health or the health of the other inhabitants of Planet Earth.

ACTIVITY 25 | What Role Do They Play?

1. Search in the library or on the Internet for information about one or more of the following topics. Be sure to investigate their relationship to the loss of stratospheric ozone.

   - Ultraviolet A
   - Ultraviolet B
   - Carbon dioxide as a refrigerant

2. Why does the ozone "hole" only last a short time? _____
   _____
   _____

ACTIVITY 26 | Debating the Ozone Issue

With the help of your teacher, set up three debate teams to tackle the following problem:

A manufacturer of aerosol spray cans for hair sprays asks government officials for permission to use CFCs until a suitable substitute can be found. If this request is not granted, his factory will close and 110 workers will lose their jobs. On hand are a group of concerned scientists, who oppose the use of CFCs for any industrial use.

One team will be the manufacturer and his assistants. Another team will be the government officials who will ask questions and render a decision. The third group will be the concerned scientists, who will provide information and evidence about the impact of CFCs on the environment.

CHAPTER 11 REVIEW | Stratospheric Ozone Loss

1. Where in the atmosphere are ozone molecules harmful? Where are they helpful?

2. How to chlorine molecules destroy ozone molecules?

3. Why is the loss of ozone in the stratosphere a global problem?

4. Explain why it may take 50 years or more to return the ozone layer to its normal thickness.

5. Why were many nations reluctant to stop using CFCs even after they were informed of their harmful effects?

WATER, WATER, IS IT FIT TO DRINK? | 12

Almost 75 percent of the surface of our planet is covered with water. Therefore, it seems that there should be plenty of clean water for everyone. In fact, sometimes there is too much water in one area and too little in another. Consider these headlines that appeared in the same paper on the same day in April 2001: "Midwest Braces for Floods" and "Florida Experiencing Its Worst Drought in 50 Years."

The word *drought* triggers a feeling of alarm in practically everyone. Water is a precious resource that is essential to life. A drought occurs when we have less water than we are accustomed to using. Because of the 2001 drought in Florida, federal approval was sought for a plan that would provide Florida with fresh water and save the state millions of dollars. The drought, along with a growing population, meant that Florida faced severe water shortages. The plan called for storing billions of gallons of untreated, polluted rainwater underground. The rainwater is contaminated with coliform bacteria, which can make people sick. Coliform bacteria are often found in sewer water. Ordinarily, that water would flow into the oceans. Depending on the amount of rain that falls, more than a million gallons a day could run off the land into the oceans surrounding southern Florida.

Normally, rainwater is captured, stored, and purified in above-ground reservoirs. This is an expensive task. To purify rainwater and store it in above-ground reservoirs would cost Florida hundreds of millions of dollars a year. A new proposal called for the **contaminated** water to be pumped into underground storage areas called **aquifers.** Aquifers normally hold fresh water that has seeped down into the aquifer from the earth's surface. In 2001, Florida's aquifers were only partially filled because of the drought. Therefore, the plan was for more than 300 wells to be built to pump the rainwater into the aquifers. In times of a water shortage, this stored rainwater could be pumped up and used for drinking and washing.

Some scientists said that the underground storage would kill the harmful coliform bacteria. Many scientists doubt that this is correct. Will underground storage really kill the harmful bacteria in the water? Will adding contaminated rainwater to the fresh water in the aquifers foul the good water that was there? Will Florida water be fit to drink? Time will tell.

Another possibility for states like Florida is to build desalination plants, which can remove the salt from ocean water. The resulting fresh water can then be stored in reservoirs.

In New Mexico and other western states, the drinking water looks—and tastes—fine. Tested chemically, though, the water is proved to have high levels of arsenic, a poison that can cause a number of different types of cancer, including cancer of the liver and kidneys. For a city or town in New Mexico to lower the level of arsenic in the water, it could cost up to $32 per home. That is why many local governments are reluctant to work on water purification.

Water, Water, Is It Fit to Drink?

Rich people can drink bottled water or have expensive filters attached to their water faucets. Poor people have no choice but to drink from the public water supply. Some people even spend a large part of their income for bottled water. If you are poor, what do you do?

Contaminated water is an increasing problem in many industrialized countries. In the next activity, you will find out what your own city or town does to purify its water.

Consider the dead zone in the Gulf of Mexico. It is called a *dead zone* because no fish or marine animals can be found there. The problem is that the water lacks dissolved oxygen, which is necessary to sustain marine life. This happened because the Mississippi River waters, which contain nutrients from fertilizers used by farmers along the Mississippi's banks, flow into the Gulf. The nitrates and other fertilizer nutrients in the runoff water feed algae in the Gulf. True, algae will produce oxygen. But, there is more to the story. The algae, being well nourished, reproduce in very large numbers. And, eventually, they die. Then, bacteria of decay in the water begin to decompose the dead algae. In the decomposition process, the bacteria use practically all of the dissolved oxygen. None is left for the fish or other animals on the floor of the Gulf. In 1998, the dead zone was the size of the state of New Jersey. Each year it grows larger.

ACTIVITY 27 | **The City Waterworks**

1. Briefly describe the geographic location of the city waterworks. _____
   _____
   _____

2. What is the source of raw water, and where is this source located? _____
   _____
   _____

3. Are there alternative sources of water? What are they, and where are they located?
   _____
   _____
   _____

4. Make a simple diagram of the flow of water from its source to the final storage tanks. Include all physical and chemical treatments of the water. (Use a separate sheet of paper.)

5. Are there any special treatments necessary before the water can be deemed fit for human consumption? If so, what are these treatments? _____
   _____
   _____

6. Are there state and/or local water quality standards? If so, what are they?
   _____
   _____
   _____

7. What methods and procedures are used to maintain water quality standards?
   _____
   _____
   _____

8. Are the facilities large enough to maintain an adequate supply of water 24 hours a day? What is the evidence? _____
   _____
   _____

Water, Water, Is It Fit to Drink?

9. In case of fire or other emergency, is there an adequate reserve of water? What is the evidence? _____
_____
_____

10. Does the facility have sufficient capacity to meet growing demands for water in the

    next year? _____   next 5 years? _____

    next 2 years? _____   next 10 years? _____

11. Can the water source provide enough raw water for growing demands in the

    next year? _____   next 5 years? _____

    next 2 years? _____   next 10 years? _____

12. What pre-service and in-service training is provided for the waterworks operators?
_____
_____
_____

13. Homeowners are charged for the water they use. Is the present waterworks the most economical for the consumer, or are there feasible methods/sources available that might reduce some of the expenses? Discuss these. _____
_____
_____
_____
_____
_____
_____

14. Do you have any suggestions as to how you or your class might help solve water resource problems? Might it be possible, for example, to sponsor a community water use awareness campaign in order to get individual homeowners to reduce unnecessary water consumption? _____
_____
_____
_____
_____
_____
_____

ACTIVITY 28 | Determining Water Consumption

During the drought of 1988 in the United States, many towns and villages had to restrict water use. Water levels fell to new lows in many places. Barges and ships even got stuck in the Mississippi River. In this activity, you will determine how much water is used in your home. You will also consider how much water could be saved if every person in your community used only the *minimum* amount required for good health and cleanliness. Then, imagine how much water could be saved if the entire nation practiced water **conservation**!

As a matter of fact, in recent years, water use in the United States has gone down. This is mainly due to more efficient means of irrigation and the use of modern toilets, which use less water than the older ones did.

Your first task is to estimate as closely as possible the daily consumption of water in your home. Below is a list of water uses and the average amount for each use. Using these figures as a guide, make an estimate of how much your own family uses daily. For example, if there are six people in your family and each takes one shower a day, that would be 150 gallons per day for showers. However, your oldest brother stays in the shower twice as long as every other family member, so it would be wiser to estimate 175 gallons for showers, or the amount used for showering by seven people. Similarly, a family of three might use only 10 gallons for washing their dishes in the sink, while your family of six might use 20 gallons or even more!

HOME WATER CONSUMPTION	
Daily Activity	**Gallons Per Use**
Bath	36
Brushing teeth	1
Clothes washer	30
Dishwasher	16
Dishwashing in sink	10–30
Shaving	1
Shower	25
Toilet flushing	1–4
Washing hands and face	2
* Water leaks	varies

* To determine the amount of water lost from a leaking faucet, do the following:

1. Place a large pot or pan under the leaking faucet.
2. Collect water for one hour.

Water, Water, Is It Fit to Drink?

3. Pour the water you collected into an empty half-gallon milk or juice container. If you have much more than half a gallon, substitute a gallon container. If you collect only a little water, use a quart or pint container.

4. Determine the approximate amount (gallons) collected during that hour.

5. Multiply by 24, and you will know how much water is lost each day.

WATER CONSUMED BY THE _____ FAMILY PER DAY

Daily Activity	Gallons Used
Baths	_____
Brushing teeth	_____
Clothes washer	_____
Dishwasher	_____
Dishwashing in sink	_____
Shaving	_____
Showers	_____
Toilet flushing	_____
Washing hands and face	_____
Water leaks	_____
Other Uses:	
Water for pets	_____
Washing car	_____
Watering lawn and garden	_____
_____	_____
_____	_____
TOTAL GALLONS PER DAY	_____

Multiply total gallons by 3.8 to convert to LITERS/DAY _____

Total Gallons/Week _____   Total Liters/Week _____

Repair leaking faucets and changing a few habits can result in the saving of a considerable amount of water over an extended period.

Water, Water, Is It Fit to Drink?

1. In which activities could less water be expended without endangering the health and welfare of you and your family? (*Example:* Use 2 gallons for flushing toilet rather than 5.)

   _____

   _____

   _____

2. How much do you estimate could be saved per week as a result of following your suggestions? _____

   _____

3. Which activities could be eliminated completely in times of severe drought?

   _____

   _____

   _____

   _____

4. How could car-washing establishments stay in business in times of water shortages?

   _____

   _____

   _____

5. How does your water-saving figure compare with the figures of other students in your class?

   _____

   _____

6. Let us imagine that 25,000 families live in your community. If each saved as much as you estimated in Question 2, how much could be saved in a week by the entire community? _____ In a year? _____

Water, Water, Is It Fit to Drink?

ACTIVITY 29 | The Sewage Treatment Plant

Your next activity is to find out what happens to the waters that end up in the sewers. What is removed? Where does the "cleaned" water go?

1. Describe the location of your local sewage treatment plant. Include in your answer how close the plant is to your city and to waterways. _____
   _____
   _____
   _____

2. What is the nature of the sewage that enters the plant? _____
   _____
   _____

3. Are there separate systems for storm runoff and raw sewage? Or are they combined?
   _____
   _____

4. What is the main disadvantage of a combined system? _____
   _____
   _____

5. Describe the treatment that the sewage gets at the plant.

   (a) What is the purpose of the primary treatment? _____
   _____
   _____

   (b) What happens to the sewage during the secondary treatment? _____
   _____
   _____

   (c) What is the tertiary treatment, if any? _____
   _____
   _____

6. On a separate piece of paper, diagram a simple flow chart, tracing the sewage from its source to its place of discharge from the plant. Indicate all physical and chemical treatments that are used.

7. What are the standards for the quality of the water discharged from the plant?

   _____
   _____
   _____

8. What is done with the sludge? In your opinion, could it be recycled? Explain your reasoning.

   _____
   _____

9. What has been done in other localities to make a profit from sludge? _____

   _____
   _____

10. What methods and procedures are used to monitor and test the efficiency of the sewage treatment plant? How often are the tests carried out? _____

    _____
    _____

11. Under what circumstances, if any, does the liquid discharged from the plant *not* meet minimum standards? _____

    _____
    _____

12. Do you believe that, as your city or town grows, the sewage treatment plant can handle the increase in raw sewage? Explain your reasoning. _____

    _____
    _____

13. What special training do the workers at the sewage treatment plant receive? _____

    _____
    _____

14. How do computers aid in the operation of the sewage treatment plant? _____

    _____
    _____

15. If the treated water is released into a river, what are the health and economic implications for the people living downstream? _____

    _____
    _____
    _____

Chapter 12 Review | Water, Water, Is It Fit to Drink?

1. Why are aquifers important? _____
   _____
   _____

2. Why must rainwater be treated chemically before it can be used? _____
   _____
   _____

3. What is an alternate source of water for coastal communities? _____
   _____
   _____

4. How did arsenic get in the drinking water of some western states? _____
   _____
   _____

5. Mention one way in which home water could be conserved. _____
   _____
   _____

THE SOLID WASTE PROBLEM: GARBAGE AND TRASH   13

Near Mexico City, tens of thousands of people of all ages live on top of a 150-acre garbage dump. Many of these human beings spend their entire lives there amid the smells and dirt. They eat the partially decayed food that is brought in mixed with other kinds of wastes. These people earn money by sifting through the solid wastes and salvaging bits of plastic and metal. Anything that can be recycled is collected and sold.

Garbage dumps in the United States are major environmental headaches. Poisons from old dumps flow into underground water systems. The poisons include paints, pesticides, acids from old batteries, and much more. The entire world has a problem dealing with solid wastes. Practically every nation in the world is running out of room to dump solid wastes.

The Fresh Kills **landfill** in Staten Island, New York, received its last bargeload of garbage on March 22, 2001. It had been receiving garbage from all parts of New York City for 53 years; originally it was scheduled to be open for only 7 years. It is 3,000 acres in size. In the 1980s, 29,000 tons of garbage were shipped there each day. If the garbage dump got much taller, it could possibly interfere with planes landing at a nearby airport. The unpleasant odor of rotting garbage was noticeable for miles.

The bad odors come from chemical compounds in the **landfill gas,** which is 60 percent methane and 40 percent carbon dioxide. (Do you recall these gases and their effect on global warming?) Landfill gas is produced when garbage decomposes. A commercial company captures and processes some of the methane gas and sells it back to New York City. However, not all the methane can be captured, so the rest is burned off in **flaring stations.** Flaring stations also burn off the odor-causing compounds. All garbage that cannot be recycled will now be shipped to Virginia and other states. The receivers of New York City garbage charge for using their landfills. In the future, the Fresh Kills landfill will be a large park with trees, basketball courts, and other amenities.

A promising solution to some of our solid waste problems is **recycling.** In addition to lessening the amount of solid waste material, every ton of recycled paper that is produced spares approximately 17 trees from being harvested. Glass bottles and metal cans can be recycled easily. Plastic bottles can't be recycled as glass bottles are, but plastic can be shredded and made into construction materials and carpet backing.

In Tokyo, the capital of Japan, residents have learned to live with garbage by recycling what they can and turning sanitary landfills into parks and other recreational areas.

The Solid Waste Problem: Garbage and Trash

Flaring station. (Photodisc CD)

Another answer is to bury the solid wastes in lined sanitary landfills. The wastes are spread in thin layers, then compacted into the smallest practical space. Finally, they are covered with soil by the end of the workday. The purpose of lining the landfill is to prevent rainwater from leaching out the toxic wastes that may be in the landfill. However, there are questions about the effectiveness of this method in preventing such leaching.

Incineration has recently become a major issue in many areas where it has been proposed or where plants have been built. As with many other solutions, there are undesirable side effects. Ashes from burned garbage contain dangerous levels of lead, cadmium, and dioxin. Lead is harmful to many organs and systems in our bodies, while cadmium and dioxin are cancer-causing agents. When the ashes are released into the air, they can also leave unsightly residues or present unknown health risks. Furthermore, not all solid waste is burnable, including the ash. Glass, metal, and old appliances do not burn. All of these nonburnables must be buried in landfills. Incineration does not end the need for landfills.

It is difficult for people to accept the disposal of their own wastes. Sometimes this reluctance is called "NIMBY." This acronym refers to an individual's attitude toward waste: Dispose of it, but "Not In My BackYard." The following activity will help you to understand the issue of solid waste disposal in your own community. More citizen awareness is essential to solving our mounting solid waste problems.

ACTIVITY 30 | Disposing of Solid Wastes

In this activity you will investigate solid waste disposal in your own community. Excellent sources of information to complete the exercise would be your town or city officials, public works department, and local newspapers. Use additional paper if necessary to answer the following questions.

1. What method(s) of solid waste disposal is used in your community?

    Burial (landfill) _____   Incineration _____

    Recycling (explain) _____
    _____

    Other (describe) _____
    _____

2. How is the waste transported to the disposal site? (Do people bring it themselves? Do trucks make deliveries? Are "transfer stations" used?) If there is an incinerator, what happens to the unburnable material and the ash? _____
    _____
    _____
    _____
    _____
    _____
    _____

3. Describe the decision-making process used in your community for determining the method of waste disposal. How are facilities, landfill sites, and so on chosen?
    _____
    _____
    _____
    _____
    _____
    _____
    _____

The Solid Waste Problem: Garbage and Trash

4. What ecological problems are associated with your community's disposal method? Answer this question with reference to animals, air, lakes, rivers, and groundwater that are affected. _____

_____

_____

_____

_____

_____

_____

_____

5. What is the per-ton cost of this disposal method? _____

6. How has this cost changed over the last five years? _____

_____

_____

7. Describe any discussions or controversies connected with future solid waste disposal in your community. What options are being considered? _____

_____

_____

_____

_____

_____

_____

_____

8. Is hazardous or toxic waste a problem? If so, describe._____

_____

_____

_____

_____

_____

_____

9. After learning about the problems and options facing your community, draw your own conclusions about the best future course for the disposal of your community's solid waste.

ACTIVITY 31 | A Pictorial Visit to a Sanitary Landfill

The sanitary landfill you are going to visit is in Oceanside, New York. It covers 187 acres, and the trash has reached a height of more than 50 meters! These pictures were taken in the summer of 1988 from a considerable distance. At that time, the landfill was operating on a limited basis; only garden wastes and building materials could be dumped. The landfill contains 45 wells that collect methane gas, which is then used to produce electricity. These wells also have another important function. They reduce the odors that come from the dump so that people can live in houses built about half a mile from the landfill. This landfill was closed in 1990.

If you were to visit this landfill today, you would see a small mountain covered with grass and trees. On top of the mountain, flaring stations can be seen as landfill gas is still being produced under the cap. It is a transfer station for garden and construction wastes (pieces of lumber, plasterboard, dead trees and shrubs, etc.). Trucks pick up the wastes and transfer them to other states.

The Oceanside dump may seem huge to you. However, it is small when compared with the Fresh Kills dump in Staten Island, New York. That landfill was discussed at the start of this chapter.

When the Oceanside landfill was opened back in 1942, lining of dumps was unknown. Lining prevents rainwater from leaching out the poisonous materials that may be buried in the landfill. An experiment was begun in 1988 at the Oceanside landfill to learn if **capping** a landfill would perform the same function as lining. Seven acres close to nearby marshland were capped. The cap was made of a sand base 30 centimeters thick, covered by a plastic liner. Another 30 centimeters of sand was deposited on top of the plastic. Finally, 15 centimeters of topsoil formed the top of the cap.

The Solid Waste Problem: Garbage and Trash

On the far left is a truck that is leaving the dump after depositing lawn clippings and garden wastes on top of the landfill. The height above ground level is more than 50 meters.

The tractor on the right is spreading clean soil over the refuse dumped that day. The four barrellike objects on the top of the landfill are vents that allow methane gas to escape or be collected for fuel to produce electricity.

The Solid Waste Problem: Garbage and Trash

The smokestacks are part of an incinerator that is no longer in operation because residents a few miles away complained vehemently about the terrible odors and smoke. The objects to the right of the incinerator are trucks filled with clean soil waiting to be unloaded.

In the last picture you see the trucks magnified. The water is a portion of a channel that connects to the Atlantic Ocean a few miles away. The marshes bordering the channel are wetlands and a hatchery and haven for young saltwater fish, clams, crabs, and other forms of marine life.

The Solid Waste Problem: Garbage and Trash

1. From 1942 to 1985, approximately 2.5 million pounds of garbage were dumped here every day the landfill was open for business.

    (a) Convert this amount to kilograms using the formula that 1 kilogram equals 2.2 pounds.

    _____

    (b) Assume the landfill was open 250 days a year. How many kilograms of garbage were deposited from 1942 to 1985? _____ kilograms

2. Why was an incinerator built next to the landfill? _____

    _____

    _____

3. Why was the landfill located in a marsh (wetlands)? _____

    _____

    _____

4. Why is this location for a landfill a poor ecological choice? _____

    _____

    _____

5. What functions are performed by the following types of heavy equipment?

    (a) bulldozers _____

    _____

    (b) cranes _____

    _____

    (c) garbage trucks _____

    _____

    (d) compactors _____

    _____

6. Should swimming and fishing be prohibited in the waters adjacent to the landfill? Justify your answer! _____

    _____

    _____

    _____

The Solid Waste Problem: Garbage and Trash

7. Instead of following the policy of the Japanese and turning former landfills into parks, baseball fields, and handball courts, should countries like the U.S. build one-family homes or even apartment buildings on former landfills to relieve housing shortages?

   _____
   _____
   _____

8. What has been the effect on the seagull and rat populations of allowing only garden and construction wastes to be dumped at Oceanside? _____

   _____
   _____
   _____

9. What happens to garden wastes and building rubbish after a dump is closed forever?

   _____
   _____
   _____

10. The land adjacent to the landfill was used for:

    marinas and boatyards   electrical power plant
    junkyards   small stores
    shopping centers   condominiums
    fire-department training center

    (a) Why is the value of nearby land increasing? _____
        _____

    (b) Why has there been a marked increase in land use between 1988 and the present?
        _____
        _____

    (c) What problems do the adjacent landowners experience as a result of the landfill?
        _____
        _____
        _____

CHAPTER 13 REVIEW | The Solid Waste Problem: Garbage and Trash

1. Why are garbage dumps said to be "environmental headaches"? _____
   _____
   _____
   _____

2. Explain why recycling is important in dealing with the garbage crisis. _____
   _____
   _____
   _____

3. What is a "lined sanitary landfill?" Why are they useful? _____
   _____
   _____
   _____

4. What is the problem with incinerating garbage? _____
   _____
   _____
   _____

5. How have e-mail and other electronic communication devices helped with the garbage crisis? _____
   _____
   _____

AFTERWORD:
SUGGESTIONS FOR FURTHER STUDY

We hope you have gained in your understanding of basic ecological concepts. Your new knowledge will help you appreciate the complexity of our world and the enormous role human beings play in it. Of course, this book does not cover all of the possible ecological consequences of human activity. We have covered some global, regional, and local problems: deforestation, the burning of fossil fuels, the search for alternative energy sources, acid deposition, destruction of stratospheric ozone, global warming, home water purification, waste water treatment, and solid waste disposal. You are encouraged to study many others.

For example, agriculture is a major topic with significant ecological effects. Can human beings continue to modify natural communities in order to produce domestic plants and animals indefinitely? What have been the effects of the introduction of chemical **herbicides** and pesticides in order to increase crop yields? What happens when natural biological diversity is replaced by cultivation of single crops over vast areas? To answer these questions, you should consider the loss of natural predators that would otherwise control insect pests, the likelihood of devastating disease, and the steady erosion of topsoil.

Another topic demanding further study is the pollution of the oceans of the world. Most of the planet's photosynthesis is carried on by marine green plants. The oceans are a source of food for people in many countries. Modern fishing methods not only wipe out species of fish, but also destroy ecosystems such as coral reefs. These practices must be stopped or controlled. There also must be a reduction in the amount of pollutants that flow into our marine ecosystems.

You could add other topics and issues—like air pollution, population growth, loss of wildlife habitats and wetlands to development, endangered species of plants and animals, and strip mining. Many, many others are also possible. Investigate some of these problems on your own. Always consider the role human beings play in these issues. Are there any solutions? What actions can people take to build healthy, sustainable ecosystems? The long-term survival of human beings depends on our answers.

ANSWER KEY

CHAPTER 1
LIVING THINGS PRESENT IN PASTURE (P. 2)
Answers may include as many of the following as possible: cattle, prairie dogs, rattle snakes, bacteria of decay, coyotes, ticks, beetles, earthworms, ants, quail, yucca plant, grass, flies, cactus, field mice, lizards, tapeworms, crows, tumbleweed, and rats.

(PP. 2–3)
- *In what ways are the cattle dependent on the grassland? Explain your answer.* Grass is the main source of food for cattle.
- *How is the grassland dependent on the cattle?* Cattle provide carbon dioxide and wastes that add nitrogen to the soil. Cattle manure interacts with soil, increasing the fertility of the grazing land.
- *How might the cattle harm the grassland?* by overgrazing
- *In what ways do the plants in the pasture compete with each other?* for light, space, and water
- *What animals might compete with the cattle for food?* rabbits, mice, prairie dogs
- *Why might this competition occur?* Plants/food might be scarce in times of drought.
- *What is the main purpose of raising cattle?* for profit
- *What might this environment look like in 10 years if humans and their cattle were removed?* Without cattle grazing, grass and other plants would grow unchecked, which in turn would provide shelter and nutrients for other animals, which in turn would attract their natural predators. Deprived of host cattle, parasites such as ticks and intestinal worms would disappear. Some insects, prey of the quail for example, would proliferate.

NONLIVING THINGS (P. 4)
What factors might influence the kinds of living things that exist in a certain area? Give three examples below:
1. temperature range
2. type of soil
3. amount of rainfall

(P. 4)
What other examples can you think of? Answers will vary.

ACTIVITY 1 (P. 5)
Individual student responses will vary.

REVIEW (P. 8)
1. the study of living creatures and their relationships to each other and to their nonliving environment
2. The nonliving parts of an ecosystem; these include chemical and physical factors.
3. amount of oxygen, temperature, amount of rainfall, type of soil, etc.
4. They study the living organisms in ecosystems such as the plants, animals, bacteria, and other decomposers.
5. Answers will vary.

CHAPTER 2
POPULATION DENSITY (P. 11)
List the two factors that increase the density of a goose population:
1. more than adequate food supply
2. lack of predators

List the two factors that decrease the density of a goose population:
1. drought, lack of food or water
2. increased number of predators, including introduction of new species

ACTIVITY 2 (PP. 12–13)
1. **Method:** Divide the picture into 4 or more sectors of equal area. Then count the plants in one of the sectors, and multiply by 4 (or whatever the total number of sectors is).

Estimation: The plant population in this photo is estimated to be about 72, although answers will vary. Draw lines to separate one sector from the next. Also, try to count *only whole plants* in the selected section.

2. **Method:** Using the estimate of 72 plants, divide 72 by 2.8m². The answer is 25.7 plants per square meter.
3. Answers will vary. To improve their accuracy, students could count all of the plants in the picture. They could also increase the size of the sample sector by doubling or tripling its original dimensions.
4. Using a larger sample sector, the size of the entire water lettuce population would be 64,250 plants (25.7 plants/m² × 2,500 m² = 64,250 plants).
5. The population density of water spiders per square meter is just over 21. To calculate the total population of water spiders for the entire swamp, you must multiply 2,500m² by 21. If the density of the water lettuce plants was affected by a high winter mortality, the effect on the spiders would be the same.

ACTIVITY 3 (P. 14)
Answers will vary with room size and attendance.

ACTIVITY 4 (PP. 15–16)
1. Quail populations could fall for lack of food. Bitter cold weather could kill weaker quail.
2. September populations of both groups are approximately the same, therefore regulated hunting does not appear to have a lasting effect on the population.

HOMEOSTASIS (P. 21)
Are humans a necessary factor in establishing a homeostatic quail population? You should be able to cite evidence to defend your response. The data show that populations in both the hunted and nonhunted areas were the same. So, humans are not necessary to establish a homeostatic quail population.

ACTIVITY 5 (PP. 24–25)
1. Answers will vary. In most cases, the rise and fall of the population will be sharp, and there will be a plateau where the population is fairly constant. All graphs should have a similar appearance.
2. As wastes accumulate, the population decreases. Other factors that increase the mortality rate include hardening of the food and competition for the food by molds and other fungi.
3. No. There was no way for the food supply to be renewed, nor could the wastes be disposed of adequately to sustain a large population.
4. Again the answer is no, for the same reasons as in the previous question, and the wild population can migrate.
5. Space is an important factor: As the natality rate goes up, the space for food and air per fly is reduced. Thus, the mortality rate will increase.
6. Trees and algae would be affected by acid rain. They also would be affected by the slow buildup of carbon dioxide in the atmosphere, which produces a warming trend. As a result, pine forests are "retreating" northward in this hemisphere.

   Deer are affected by soil and water pollution as well as by human beings. The loss of the ozone layer will produce skin cancers in animals as well as people. As our biosphere becomes more and more polluted, the mortality rate will increase for many forms of life.

REVIEW (P. 26)
1. natality, mortality, immigration, emigration
2. The population will decrease.
3. Because the data will be more accurate and significant.
4. The organism populations in an ecosystem are stable; there is variation from year to year, but generally the numbers remain fairly constant.
5. The Canada geese population will probably decrease slightly because bald eagles are not active predators. They eat food in garbage bins and steal food from smaller, weaker birds including ospreys. Only as a last resort will the eagle become an active predator.

CHAPTER 3
COMMUNITIES (P. 30)
Note below the similarities you believe exist among these five photographs. The photographs all contain plants of varying sizes. In some, you can also see animals. It should be assumed that animals are present

in all the photographs even though they may not be visible. Air, soil, and water are present (even in the desert community).

*All five photographs represent what could be called **natural biological communities**. From your observations, how would you define a natural biological community? Write your best definition below.* A natural biological community is any natural relationship involving plants and animals and their interactions with each other.

ECOSYSTEMS (P. 33)

A list of abiotic factors might include: light, minerals, wind, humidity, elevation, gravity, the medium upon which the organisms exist (water, mud, sand, rock), predominant land forms, wave action, and tides.

ACTIVITY 6 (PP. 36–37)

Student findings will vary.

The main abiotic factors influencing these two communities are moisture content and soil temperature. Hours of exposure to sunlight would be another factor. Also important are the amount of decaying material (humus) in the soil and the percent of sand and clay. Students should discover that the higher the moisture and humus content of the soil, the greater the kinds and numbers of animals present.

REVIEW (P. 38)

1. In a community, living populations interact with each other. An ecosystem comprises both the living and nonliving factors in an environment.
2. Wide variations in abiotic factors, such as sunlight, water, temperature, have a profound effect on the living communities that can thrive in a specific environment, for example in a desert or in a bog.
3. Among desert predators are hawks, vultures, roadrunners, and owls.
4. The spines of cacti are actually leaves; photosynthesis is carried out in the fleshy green stems. Spines are adapted to deter predators.
5. Prairie grasslands thrive in subhumid to humid climates; soil conditions are a significant factor.

CHAPTER 4
FOREST TYPES (P. 39)

How many centimeters of rain fell last year where you live? Answers will vary.

CAUSES OF DEFORESTATION IN THE TROPICS (P. 41)

"slash-and-burn agriculture." What does this term mean? The trees are cleared with a machete or other cutting tool. Then the small ground plants and stems that are still standing are burned.

(P. 41)

As the trees fall, what will happen to the younger trees nearby? The younger, smaller trees in the path of the large tree will be knocked down. Most will either be badly damaged or will die.

DEFORESTATION AND FARMING IN THE UNITED STATES (P. 42)

None of the trees that were here when the Pilgrims arrived are alive. Why not? Most trees don't live 350 years.

(PP. 42–43)

1. over 700 million acres
2. about 120 million acres
3. There was about $1\frac{1}{2}$ times as much acreage devoted to forests.
4. The reasons are: better farming methods, use of machinery rather than animals, farms were started in the Midwestern prairies (although no new farms were started in the Northeast). Midwestern farms were easier to work than farms on mountainsides. Wood was no longer used for heating and cooking.
5. coal, petroleum products, and natural gas
6. Plastic has replaced wood in many applications, preserving forests.
7. Computers and e-mail reduce reliance on paper, preserving forests.

REASONS FOR PRESERVING FORESTS (P. 44)

Laws for protecting the forests must be enforced by the nations of the world. Do you think these measures would work? Explain your answer. Answers will vary.

ACTIVITY 7 (P. 45)

Student responses will vary.

ACTIVITY 8 (P. 46)

Student responses will vary.

REVIEW (P. 48)

1. for firewood, for lumber for homes and furniture, to turn the forest land into farms, to have more living space
2. to convert the wet lands into dry land for homes; to use the water for homes and industrial and commercial purposes
3. Forests preserve the climate, provide homes for wildlife, prevent floods, are a source of beauty, provide medicines and drugs, lower the amount of carbon dioxide in the atmosphere, add oxygen to the atmosphere, and remove pollutants from the soil and atmosphere.
4. Wetlands are a home for wildlife. They also prevent floods and can be enjoyed by people.
5. Write to local, state, and federal officials stressing your feelings about preserving these national and international treasures.

CHAPTER 5
(P. 49)

What activity is it? Obtaining food.

ROLES WITHIN THE FOOD CHAIN (P. 50)

What organisms other than those shown might actually exist in this food web? There are many, but here are a few: bison, antelope, and wolf.

What would probably happen to this food web if the grass on the prairie were suddenly to disappear? Since the grasses are the producers, the first-order, second-order, and third-order consumers would either leave or die off.

What would happen if everything remained the same except that the decomposers disappeared? Nothing immediately, but in a few years the food web would die out, because there would be no replenishment of the minerals in the soil and the grasses (producers) could not grow.

ACTIVITY 9

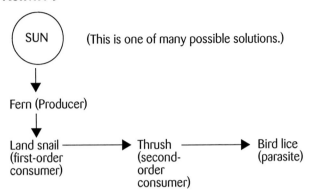

(This is one of many possible solutions.)

ACTIVITY 10

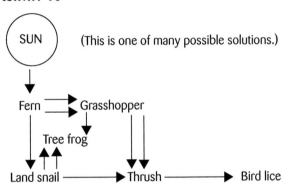

(This is one of many possible solutions.)

ACTIVITY 11

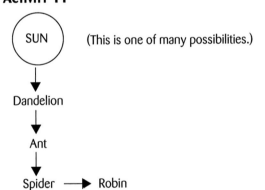

(This is one of many possibilities.)

How does the food chain differ from the one you drew before? The consumers are smaller in size. The producers are mainly smaller in size and fewer in number.

How do humans affect the ecology of a city area? Destruction of the topsoil and its replacement with blacktop and concrete have eliminated many producers. Without producers, no consumers can exist. (Parks and vacant lots are partial exceptions because they are often tended by humans.) The elimination of weeds and mosquitoes are two examples.

ACTIVITY 12

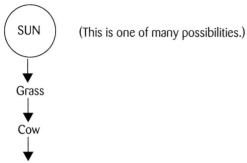

(This is one of many possibilities.)

Activity 13 (p. 65)

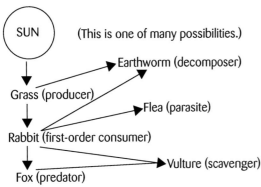

Activity 14 (p. 66)
Answers will vary.

Review (p. 67)
1. A food chain is a simple straight-line process going from producer to first-, second-, and possibly third-order consumers, and ending with the decomposers. In a food web, there are a number of second- and third-order consumers.
2. There are few elephants because many producers (plants) are necessary to support one animal the size of an elephant.
3. In mutualism, both organisms benefit. In commensalism, one organism benefits; the other is unaffected positively or negatively.
4. A lichen consists of a fungus and algae, living in a mutual arrangement.
5. Food niche means the place a particular organism has in its ecosystem in terms of food. There are herbivore food niches and carnivore food niches.

CHAPTER 6
The Carbon Cycle (p. 71)
1. volcanoes, burning of fuels, plant and animal respiration
2. atmosphere, soil, water
3. Photosynthesis could not take place.
4. Path 1: Marine plant (photosynthesis), marine animal, to ocean, to atmosphere
   Path 2: Marine plant (photosynthesis), marine animal, organic sediments, oil, power plant, atmosphere
5. Only green plants are able to remove carbon dioxide and convert it to organic molecules, which then become carbon dioxide again and enter the atmosphere.
6. Yes. Plants and trees would continue to be involved in the carbon cycle.

The Water Cycle (pp. 72–73)
1. Condensation (conversion of vapor to liquid) and evaporation (conversion of liquid to vapor) are opposites.
2. Sometimes rain evaporates before reaching the ground.
3. The water table is the level below which all of the ground is saturated with water.
4. Infiltration is snow or rain that seeps into the ground.
5. Humans and animals are missing. They add water to the atmosphere, the ground, and bodies of water. They also remove water from fresh-water sources.

Activity 15 (p. 73)
Students should see fine particles of water. This shows that respiration by animals produces water vapor, which can then enter the atmosphere.

The Phosphorus Cycle (p. 74)
1. The soil is the main sink.
2. They get phosphorus from eating plants or animals that have eaten green plants.
3. practically none
4. Runoff is material that ends up in any body of water. It may have been absorbed first by the soil, or it may have just run off the land into the water.

(p. 75)
1. Trees, plants animals, water, and decomposers can be found in all three.
2. Fungi, worms, and bacteria are involved in decomposition.
3. water and phosphorus
4. carbon dioxide

Activity 16 (p. 77)
Student responses will vary for predicting the snail's lifespan and for the data table.

Questions (p. 77)

1. Droplets of moisture on the inside of the glass container are evidence of condensation.
2. The greenhouse effect would kill the snail and possibly the plant.
3. All three cycles are going on. Carbon cycle: Plant makes food by photosynthesis; plant and snail carry on respiration. Water cycle: Both use and excrete water. Phosphorus cycle: Both organisms use and excrete phosphorus.
4. The plant provides food and oxygen.
5. The snail provides carbon dioxide and phosphorus.
6. Bacteria are present.

Review (p. 78)

1. Carbon atoms are needed to make practically all biochemical compounds.
2. The sun provides the energy that drives the cycle.
3. The sun causes evaporation due to its heat energy.
4. Only tiny amounts of new atoms enter our closed ecosystem from space each year. Therefore, the earth's atoms must be used (and recycled) most efficiently.
5. Decomposers break down organic molecules to inorganic elements and compounds so that they can be reused to form new compounds.

CHAPTER 7
Activity 17 (pp. 84–85)

1. The pond is in its old age. The observations for this conclusion include the facts that (a) there is very little water and (b) land plants have invaded from the shore.
2. In 10 years there will probably be very little water in the pond. More and more shore organisms, both plant and animal, will invade the space that currently is occupied by the pond. In 100 years the pond will be gone, filled in with organic matter. Only land organisms will live in the area now covered by the pond.
3. Water snails may be replaced by land snails. As the pond fills in and the water disappears, there will be less water for the insects, so fewer insects can live in the pond. Fewer insects and less water will mean fewer frogs and salamanders and red-winged blackbirds, which feed on the insects. Fewer frogs and salamanders plus less water will reduce the heron and water-snake populations. Ultimately, these animals will be replaced by land organisms such as rabbits and mice.
4. • Remove some of the plants manually.
   • Destroy the excess water plants chemically.
   • Change the pH of the water.

Activity 18 (pp. 87–93)

Use the records in Figure AO to calculate the population densities of the canopy trees.

Shagbark hickory: 0.01 per meter^2 (1 tree per 100 meter^2)

Yellow chestnut oak: 0.02 per meter^2 (2 trees per 100 meter^2)

Red oak: 0.03 per meter^3 (3 trees per 100 meter^2)

Figure AA	How Many	Percent of Total (%)
Sassafras	5	71.4
Persimmon	2	28.6
TOTAL	7	100.0

Figure AB	How Many	Percent of Total (%)
Winged Elm	1	11.1
Persimmon	3	33.3
Sassafras	5	55.6
TOTAL	9	100.0

Figure AC	How Many	Percent of Total (%)
Winged Elm	20	34.5
Persimmon	22	37.9
Sassafras	9	15.5
Black Cherry	3	5.2
White Ash	4	6.9
TOTAL	58	100.0

Figure AD	How Many	Percent of Total (%)
Black Cherry	1	7.7
White Ash	2	15.3
Persimmon	3	23.1
Winged Elm	1	7.7
Bitternut Hickory	3	23.1
White Oak	3	23.1
TOTAL	13	100.0

Figure AE	How Many	Percent of Total (%)
Bitternut Hickory	4	22.2
Shagbark Hickory	2	11.1
White Oak	6	33.3
Yellow Chestnut Oak	3	16.7
Red Oak	3	16.7
TOTAL	18	100.0

Specific Tasks for This Problem (pp. 94–96)

1. Since the land was under cultivation for the five previous years, any seeds from trees had to be brought there by the wind or animals. A few years are necessary for the seeds to germinate and grow into seedlings.
2. There are more trees in the understory. Understory trees require less soil and space than canopy trees. Thus, more understory trees can grow in a quadrat of the same size.
3. a. Winged elm, persimmon, sassafras, black cherry, and white ash disappeared.
   b. Bitternut hickory, shagbark hickory, white oak, yellow chestnut oak, and red oak appeared.
   c. Hickory and oak trees replaced all the other trees.
4. Black cherry, white ash, persimmon, winged elm, bitternut hickory, and white oak are present at 50 years.
   Prediction: All trees present at 50 years will be present at 100 years.
   Reason: This is a climax forest and it will remain unchanged. Barring any catastrophe such as fire, it will remain the same indefinitely.
5. At five years, there were no trees, only seedlings. The presence of trees that provide shade in the 50-year quadrat would account for the cooler air and soil temperature.
6. *Predicted Climax Forest Type:* Oak-hickory.
   Reason: Oaks and hickories are present in the canopy and in the understory; beeches and maples are not growing in either portion of the forest.
7. *Prediction:* no
   Reason: These trees are present in both the canopy and understory when the forest is 20 years old, but they are missing at 50 years.
8. There are fewer trees in the Figure AO canopy. Since the trees in Figure AO are 100 years old (twice the age of the trees in Figure AD), the 100-year-old trees probably have longer limbs, with more leaves that will shield more of the forest floor. Without light, younger trees of the climax type could not grow. Thus, there will be fewer but larger trees in the 100-year-old forest.
9. The first trees to gain a foothold are the persimmons, sassafras, and elms. They make the environment favorable for the growth of ash and cherry trees. Finally, the climax trees appear in the forest. These are the oaks and hickories.

Activity 19 (p. 97)
Answers will vary.

Review (p. 98)
1. They are changed by succession, natural events, and the efforts of humans.
2. Communities are destroyed by fires, lumbering, or clearing of the land for homes or industrial purposes.
3. The roots of trees, shrubs, and grasses hold the soil in place.
4. They are the first to grow on bare rock.
5. Because a community can change owing to many factors, including those mentioned in Answer 2.

CHAPTER 8
CARRYING CAPACITY (P. 100)
List the factors other than space that affect human carrying capacity. water, temperature, altitude

SMOG (P. 102)
List three additional things you and your family could do to reduce air pollution. You could ride your bike or use public transportation rather than using the car; drive less; avoid having the car engine idling more than 30 seconds; keep the engine tuned (especially the spark plugs); use the air conditioner fewer hours each day.

ACTIVITY 20 (P. 104)
In which state did rain with a pH of 4.2 fall? Ohio
Find the state of New York. What was its lowest pH? 4.4
Locate Massachusetts. What was its lowest pH? 4.6
Find your state. What was its lowest pH? Answers will vary.
How do pH readings for rain in the West and Midwest compare with those in the Northeast? Most of the readings in the Midwest are above 5.3. Some pH readings are as high as 5.6. In the Northeast, all of the readings are around 4.6 or lower.
Which state(s) had no acid rain (normal rain has a pH of 5.5)? Washington, Oregon, Montana, North Dakota, Minnesota, Nevada, Idaho

ACTIVITY 21 (P. 107)
Answers will vary.

REVIEW (P. 108)
1. Nitrogen oxides and sulfur dioxide are major air pollutants, resulting from burning coal and oil; from oil refineries and pulp and paper mills; automotive emissions and electrical power generation.
2. Sulfur dioxide is the major contributor to regional haze. Smog is caused by nitrogen oxides reacting with oxygen in strong sunlight to produce ozone.
3. The westerlies (winds) blow acid deposition toward the East Coast.
4. It is expensive and time-consuming to remove pollutants.
5. Water and food are in short supply in the desert, and the temperature extremes are unfavorable for most people.

CHAPTER 9
(PP. 109–110)
What is that temperature in degrees Fahrenheit? 57 °F.
In what way is the x-axis unusual? The zero mark is not at the bottom of the x-axis.
What was the temperature deviation in the year 1900? Almost zero. *What does this mean?* It means that the temperature was practically normal on a worldwide basis.
Approximately when did the warming trend begin? 1935
What were the three coldest years in the twentieth century? 1904, 1908, 1918
Compare the temperature deviation of the 1950s to that of the 1980s. In the 1950s, the average temperature deviated from the norm very little. In the 1980s, the average temperature deviation was approximately +0.03 degrees Celsius.

THE EFFECTS OF GLOBAL WARMING (P. 112)
In the spaces provided, name three diseases that are transmitted by insect bites. malaria, dengue fever, and Lyme disease

Ecosystem	Change
Tundra	Melting of the permafrost, glaciers, and ice cap; changes in plant and animal life (fewer conifers, extinction of polar bears, etc.)
Desert	Less rainfall or more rainfall, affecting living organisms there
Mountain regions	Loss of snow at the peaks; changes in trees, ground plants, and animal life
Wetlands	Due to hot temperatures, drying will be speeded up; change in biotic nature of the wetlands

CAUSES OF GLOBAL WARMING (P. 114)
1. burning of fossil fuels, plant and animal respiration, decomposition
2. the burning of fossil fuels, decomposition to some extent

3. They remove carbon dioxide from the atmosphere.
4. More carbon dioxide is released from its sources than is taken in by the sinks.

(P. 115)

Which greenhouse gas is most prevalent in the atmosphere today? carbon dioxide

Using the data in the table, explain why carbon dioxide is the most important greenhouse gas. It is the most prevalent and can be controlled to some extent. Some industrialized countries are now limiting their output.

THE GREENHOUSE EFFECT (P. 117)

Now it is your turn to think of solutions. Answers will vary.

ACTIVITY 22 (PP. 119–120)
Analysis
1. Answers will vary.
2. One bottle had more holes than the other, allowing it to cool.
3. The plastic wall represents the greenhouse gases.
4. the one with seven holes
5. The wind entering through the holes would cool the air inside the bottles.

REVIEW (P. 120)
1. It will lead to serious health problems, changes in ecosystems, a rise in the sea level, and changes in rainfall amounts.
2. There would less carbon dioxide and black soot aerosol released into the atmosphere.
3. The greenhouse effect is the process by which certain gases in the atmosphere prevent the escape of infrared radiation into outer space.
4. Carbon dioxide emissions can be reduced by the following methods: Plant more trees that use carbon dioxide for photosynthesis. Legislate further reductions in fossil fuel burning. Develop alternative energy sources and technologies; solar energy, for example, produces no carbon dioxide.
5. *Sources* = respiration, decomposition, deforestation, burning of fossil fuels, plowed fields, erosion of carbonate rocks.

Sinks = Green plants (particularly tropical forests), the ocean, sea creatures with shells, coal, peat.

CHAPTER 10
(PP. 121–122)

If a sports utility vehicle (SUV) gets 15 miles to the gallon, how many gallons would it need to travel 750 miles? 50 gallons

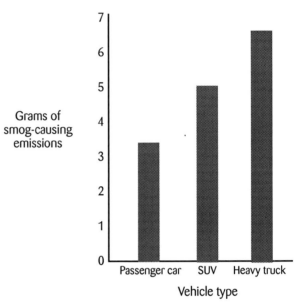

How many grams of smog-causing pollutants would it release during that trip? 3900 g

How many kilograms of smog-causing emissions would be released by American vehicles into the atmosphere this year? 17,600,000 kg

HYBRID AUTOMOBILES (P. 123)

Similarities	Differences
Both have 12-volt batteries and engine shut-off.	Mild hybrid has 42-volt battery pack, full hybrid has 300-volt battery pack.
Battery packs recharge when braking.	At low speeds, the full hybrid is powered only by the battery pack.
Both run on gasoline and battery power.	Mild hybrid has 6-cylinder engine, full has a 4-cylinder.

ANOTHER SOURCE OF ENERGY (P. 125)

How do think this car is powered? solar energy

ACTIVITY 23 (PP. 126–127)
Student responses will vary.

ACTIVITY 24 (PP. 128–129)
Student responses will vary.

REVIEW (P. 131)
1. Calling them light trucks keeps down the cost. Manufacturers can sell more of them at a lower price.
2. It saves fuel and lowers pollution.
3. Hybrids reduce the smog-causing emissions.
4. The disadvantages of battery-powered cars still outweigh the advantages.
5. Answers will vary.

CHAPTER 11
(P. 133)

Oxygen	Ozone
colorless	*a blue gas*
odorless	Has a strong odor
Composed of two atoms of oxygen	Composed of three atoms of oxygen
Has no unusual function	*In the stratosphere, it absorbs dangerous solar ultraviolet radiation.*
Is essential for most life forms	*Poisonous to living plants and animals close to the earth's surface*

OZONE LOSS (PP. 136–138)
How many months does the "hole" last? approximately two months

During which month is the ozone loss the greatest? October

Where are the ozone molecules found, above or below the line in the graph? below the line

Explain your reasoning. The depth of ozone molecules (in km) is depicted by the line.

Look at the years from 1960 to 1970 on graphs 2 and 3. How do they compare? For these years, the lines on the graphs are going in opposite directions. The line on Graph 2 is declining, while the line on Graph 3 is rising.

What possible conclusion can you reach from looking at these two graphs? As the CFCs increase, the amount of ozone in the atmosphere decreases.

Examine Graph 2 again. What effect did this release of CFCs have on the amount of ozone over the Antarctic in October of the early 1980s? Around 1984, the amount of ozone, which had been increasing in the stratosphere, began to drop sharply.

The worldwide use of CFCs was increasing at a rate of 3 percent per year starting in 1985. What evidence for that increase do you see in Graph 2? A marked, yearly decrease in ozone in the stratosphere continuing until 1992.

In Graph 3? The relative increase of CFCs in the stratosphere reaches an apex in about 1986, reflecting the worldwide increase in use in 1985.

[1987 Montreal agreement]: How is this reflected in Graph 3? Around 1987, the relative increase in CFCs in the stratosphere slowed, which can be attributed to the ban on CFCs.

What was the effect on the ozone over Antarctica as shown in Graph 2? How can you explain this from what you have read so far? Because ozone levels continued to drop, it is probably because CFCs last for about 100 years before they break down into other substances.

ACTIVITY 25 (P. 139)
Student responses will vary: polar meteorology, climate, temperature, for example.

ACTIVITY 26 (P. 139)
Student responses will vary.

REVIEW (P. 140)
1. Ozone is harmful in the lower regions of the atmosphere (troposphere); it is helpful in the stratosphere, where it occurs naturally.
2. They are involved in a chain reaction in which the chlorine atom is used over and over again; this process destroys ozone molecules.
3. It's a problem because the ozone layer is over the entire planet, not just over one area or country. Thinning of the ozone layer makes everything on Earth susceptible to harmful radiation from space.
4. This is because CFC molecules remain in the stratosphere for a long period of time—up to a century.
5. Many industries were dependent upon CFCs. For example, CFCs are used as refrigerants and for aerosol sprays. In the developing countries

of the world, CFCs were important for trade—in the export of manufactured materials to other countries.

CHAPTER 12
ACTIVITY 27 (PP. 143–144)
Student responses will vary.

ACTIVITY 28 (P. 147)
Student responses will vary.

ACTIVITY 29 (P. 148-149)
Student responses will vary.

REVIEW (P. 150)
1. Aquifers are large, natural underground storage areas for water. The water can be used in homes and businesses. The water can be removed by wells.
2. Rainwater may be contaminated with harmful bacteria.
3. An alternative is desalinization—using solar energy to remove the salt from ocean water.
4. The arsenic is normally found in the volcanic soil in this region.
5. Options include the following: fix leaky faucets; take shorter showers; avoid running water for unnecessary reasons.

CHAPTER 13
ACTIVITY 30 (PP. 153–155)
Student responses will vary.

ACTIVITY 31 (PP. 159–160)
1. a. 1.136 million kg
   b. 12.496 billion kg (1.136 million kg/d × 250 d/y × 44y)
2. The incinerator uses the methane gas produced in the landfill to get rid of burnable wastes.
3. It was located here to keep the odors and ugly view of garbage away from homes and people.
4. It was a poor choice because poisonous wastes in the garbage could leach out into the waters and kill off the wetland inhabitants (fish and birds, crabs, etc.).
5. a. Bulldozers move mounds of garbage.
   b. Cranes lift the garbage out of, or into, trucks.
   c. Garbage trucks carry the garbage to or from the landfill.
   d. Compactors compress the garbage so that it takes up as little space as possible.
6. Yes, because the leaching out of toxic materials from the landfill could contaminate the water.
7. No, for the same safety reasons as cited in Answer 6. Also, landfill gases may escape from under the cap.
8. These populations have been reduced.
9. Recycle what can be used; compost the garden wastes. Ship out the rest to another landfill.
10. a. The land value is not increasing. People know there is a dump close at hand and do not wish to live close to it.
    b. Immigration is the main reason.
    c. Property values are low. Even capped with grass and trees, the dump area looks strange.

REVIEW (P. 161)
1. One problem is that of space; there isn't enough room to handle all our garbage. The other problem is that poisonous materials and acids get into the soil and nearby waters.
2. Reusing as many materials as possible reduces the amount of garbage that has to be disposed of.
3. A plastic lining is at the bottom of the landfill. Wastes are spread in thin layers on top of the lining. Then they are compacted and covered with soil. They are believed to prevent leaching of the harmful materials out of the landfill.
4. Ashes may contain harmful heavy metals such as lead. Not all wastes can be burned; glass is one example.
5. The need for paper has theoretically been reduced, thus reducing the amount of waste paper in our garbage.

APPENDIX 1
(P. 177)
1. 0.2375 liters; 237.5 milliliters
2. 101.6 centimeters; 1016 millimeters
3. 1350 grams
4. 42.8 degrees F
5. 28.35 inches (rounded off)
6. 4.24 quarts
7. 25.56 degrees C (rounded off)
8. 7.6 liters

APPENDIX 1

METRIC CONVERSION DATA

Scientists the world over use the metric system of measurement. This is different from the English system, with which many of us have been familiar. The following information may be helpful to you while participating in some of the activities contained in these pages.

Measurement of Length

Metric	English
1 kilometer (km) = 1000 meters = 0.621 miles	1 mile = 1.609 km
1 meter (m) = .001 km = 39.37 inches = 3.28 feet	1 foot = 0.305 m
1 centimeter (cm) = .01 m = 0.3937 inches	1 inch = 2.54 cm
1 millimeter (mm) = .001 m = 0.0394 inches	1 inch = 25.4 mm

Measurement of Volume

Metric	English
1 liter (l) = 1.06 quarts	1 quart = 0.95 liters
1 millimeter (ml) = .001 liter	1 quart = 950 ml
	1 gallon = 3.8 liters
	1 pint = 475 ml

Measurement of Mass

Metric	English
1 kilogram (kg) = 1000 grams (g) = 2.20 pounds	1 pound = 0.45 kg
	1 pound = 450 grams
	1 ounce = 28 grams

Measurement of Temperature

	Metric—Celsius (C)	English—Fahrenheit (F)
Water freezes	0 degrees C	32 degrees F
Water boils	100 degrees C	212 degrees F
Number of divisions between these 2 points	100	180

Therefore,
1 degree C = 1.8 degrees F

Appendix 1

To change from Fahrenheit to Celsius: degrees C = $\frac{\text{(degrees F} - 32)}{1.8}$

To change from Celsius to Fahrenheit: degrees F = (degrees C × 1.8) + 32

Examples: (a) 98.6 degrees F (human body temperature) = _____ degrees C

$$C = \frac{(98.6 - 32)}{1.8} \qquad C = 37 \text{ degrees}$$

(b) 20 degrees C = _____ degrees F

F = (20 × 1.8) + 32

F = 68 degrees

You might want to try some of the following conversions to check your own ability.

1. One-half pint = _____ liters = _____ milliliters
2. 40 inches = _____ centimeters = _____ millimeters
3. 3 pounds = _____ grams
4. 5 degrees C = _____ degrees F
5. 720 millimeters = _____ inches
6. 4 liters = _____ quarts
7. 78 degrees F = _____ degrees C
8. 2 gallons = _____ liters

Grid Paper

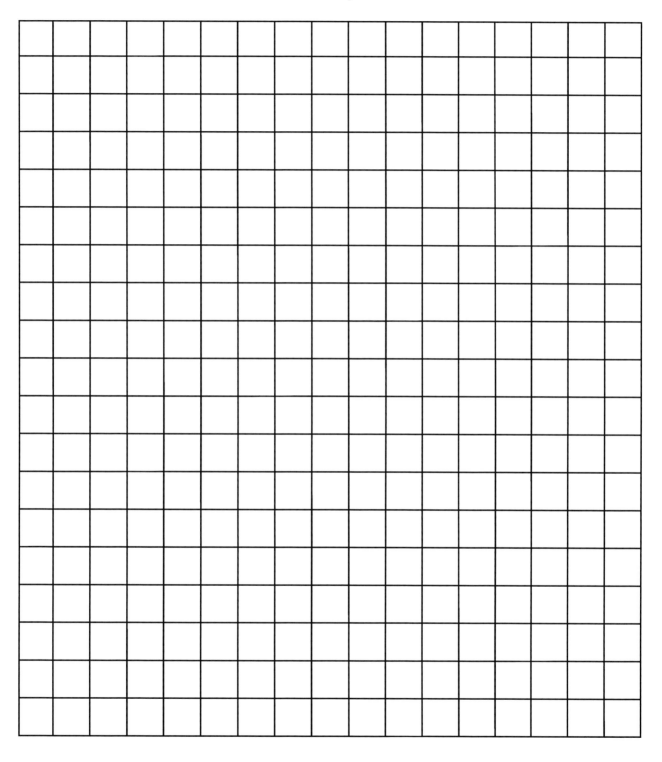

GLOSSARY

A

abiotic — Pertaining to nonliving factors in the environment. Temperature and rainfall are two examples.

acid deposition — Acid rain, snow, or solid particles that contain elevated concentrations of sulfuric or nitric acid.

acid precipitation — Rain or other forms of precipitation containing elevated levels of sulfuric or nitric acid.

active predator — A consumer that feeds only upon prey it has caught.

aerosol — Droplets or particles suspended in the atmosphere, emitted naturally (for example, in volcanic eruptions) and as the result of human activities such as burning fossil fuels.

agriculture — The production of crops and livestock.

algae — Primitive green plants (usually water plants) capable of making their own food.

aquifer — A zone under the earth's surface that holds water.

atmosphere — The gases that surround our earth.

B

biotic — Pertaining to living factors in the environment.

brackish — Slightly salty (used to describe water).

C

canopy — The uppermost branchy layer of a forest. During the growing season the leaves on trees in the canopy receive the greatest share of sunlight, while shading the forest floor.

capping — Covering a garbage dump site with plastic and soil to prevent water from leaching toxic chemicals out of the dump.

Glossary

carbon cycle	Circulation of carbon atoms through the atmosphere, biosphere, hydrosphere, and lithosphere.
carnivore	Animal that feeds or preys on other animals.
carrying capacity	The number of organisms that can be supported by a habitat or by our entire planet.
catalytic converter	A device added to the exhaust systems of motor vehicles to cut harmful emissions.
catalyst	A substance that changes the speed or yield of a chemical reaction without itself being consumed or chemically changed.
chlorofluorocarbons (CFCs)	A class of organic chemicals that destroys the ozone layer in the upper atmosphere.
chlorophyll	The green pigment found in plants that is capable of forming carbohydrates from water and carbon dioxide using light energy.
closed ecosystem	A self-sufficient working unit of living and nonliving components.
climax community	A complex community that tends to remain stable over a long period of time; the last community formed during succession.
colonize	To establish or form a colony; a colony is a group of animals or plants of the same kind of living thing living together.
commensalism	A natural relationship between individuals in which only one member of the pair benefits, while the other is neither benefited nor harmed.
community	In ecology, a group of interacting populations in time and space. Sometimes a particular subgroup may be specified, such as the fish community in a lake or the soil arthropod community in a forest.
competition	Rivalry between organisms of the same species (intraspecific) or between organisms of different species (interspecific) for biotic or abiotic factors such as food or space.
condensation	Liquid created by cooling vapor.
coniferous	A term used to describe trees that produce their seeds in cones. Evergreens are examples of coniferous trees.
conservation	The wise use of natural resources; human behavior that guards and preserves the resources of the earth.
consumer	An animal or plant that feeds upon a producer (green plant) or another consumer.

Glossary

contaminate	To make impure or to infect; to pollute.

D

deciduous	A term used to describe broad-leafed trees that lose their leaves during the fall.
decomposer	A primitive plant that lacks chlorophyll. It digests and breaks down the wastes and dead bodies of plants and animals into simpler chemicals. The simple chemicals are used both by the decomposers and by more complex plants to carry on their own life processes.
deforested	Removal of forest areas for logging or agricultural purposes.
density	The number of individuals of a particular species per unit area at a particular time.
dependence	Relying on something else for support, food, or some other necessity.
drought	A period of time during which there is insufficient rainfall.
dry tropical forest	Small forests in the Tropics that receive little rainfall.
dung	Manure; solid digestive wastes from an animal. Also called scat.
dynamic	Relating to or tending toward change. All populations exist in a dynamic state. Population density usually varies in cycles.

E

ecologist	A scientist who studies the interactions of living and nonliving factors in the environment.
ecology	The science that studies the relationships and interactions that exist between living things.
ecosystem	All of the biotic and abiotic interactions that occur in a natural community.
emigration	Departure of an organism from a particular population.
emissions	Pollution discharged into the atmosphere from industrial facilities; smokestacks or other vents, from machine exhausts, and from residential chimneys.
endangered species	A species of plant or animal that is close to extinction, usually due to few offspring being born.
energy	The capacity to do work. Light and heat are forms of energy.

Glossary

environment	The surroundings that affect the life and development of organisms.
erosion	The wearing away of rock or soil by natural forces such as water and wind.
experiment	A controlled scientific test undertaken to discover new information or to study some questionable fact or concept.
evaporation	Vapor created by heating liquid.
extinct	No longer living (pertains to an entire species that has died out).
extinction	A condition in which no members of a species are alive any longer.

F

flaring station	A pipe-like device that burns off landfill gas when ignited.
food chain	The flow of energy from one organism to another.
food pyramid	A concept describing the amount of energy that passes from one level to the next in a food chain or web. There is always a loss of 10 to 20 percent of the energy as you go from one level to the next higher level.
food web	The exchange of energy between organisms; more complex than a food chain. A food web can be considered to be a set of interrelated food chains.
fossil fuels	Coal, petroleum products, and natural gas.
fuel cell	A device that produces electricity by chemical reactions.
fungi	Multicellular organisms that cannot make their own food and instead acquire nutrients by absorption. Molds and mushrooms are examples of fungi.
fungicide	A chemical that kills fungi.

G

germinate	To begin embryo growth within a seed.
global warming	Rise in the near-surface temperature of the earth, primarily owing to increased emissions of greenhouse gases.
grazing	Usually refers to animals that feed directly on grasses and the plants associated with them.
greenhouse effect	The trapping of the sun's heat by gases in the atmosphere.

greenhouse gases	Mainly carbon dioxide, methane, and oxides of nitrogen.

H

habitat	The place or environment in which an organism lives.
herbaceous	The condition in which a plant has soft stems rather than hard woody stems.
herbicide	A chemical that kills green plants. Some herbicides are specific in that they kill only specific plants.
herbivore	An animal that feeds solely on plants.
homeostasis	A natural occurrence during which an individual, a population, or an entire ecosystem regulates itself against negative factors and maintains an overall stable condition.
hormones	Chemical messengers transmitted in animal body fluids or the cell sap of plants. Hormones affect many life activities such as growth and reproduction.
host	An organism on which a parasite depends for food.
hybrid car	A car that runs on both gasoline and battery power.
hypothesis	A statement that is temporarily adopted to explain certain facts. Its validity is investigated by experimentation.

I

immigration	Movement of an organism into a new area.
incineration	Disposal of solid waste by burning.
inorganic	Non-carbon-based minerals.
insecticide	A chemical that kills insects.
interaction	A relationship in which one thing affects another and produces a change in the object or thing affected.
irrigation	Supplying water to meet needs of plants.

L

landfill	A site used for burial of solid waste.

Glossary

landfill gas	Composed of methane and carbon dioxide. It is formed when garbage decays.
larva	The immature, wingless, wormlike form that hatches from the eggs of certain insects. Plural form is larvae.
limited resource	A reusable resource such as water. When more is used than is returned to the environment, that resource becomes limited.
lipid	Large organic molecules that don't dissolve in water.
locomotion	The ability to move about.

M

malignant	Deadly.
mangrove	A type of tree that grows only in swampy places.
metamorphosis	The process of changing shape or form.
migrate	Move from one geographical region to another.
mortality	Death rate of a population.
mutualism	A relationship between two organisms in which both organisms benefit.

N

natality	Birthrate of a population.
natural biological community	See *community*.
niche	The role that an organism plays in a community. Niches are often associated with food-getting.
nutrient	A substance found in food that is essential for the health and growth of an organism.
nutrient cycle	Circulation of nutrients through an ecosystem.

O

optimum	The best or most favorable conditions. Optimum (or *optimal*) conditions favor the survival of individuals and, in turn, populations and ecosystems.

Glossary

organic	Relating to those chemical compounds that are found exclusively in living things. Also widely used to describe foods grown or raised without artificial chemicals.
organism	A living thing.
overpopulation	The condition that results when there are too many individuals in a population for the ecosystem to support.
ozone	In the stratosphere, ozone is a natural form of oxygen that shields the earth from ultraviolet radiation. Closer to Earth, in the troposphere, ozone acts as a chemical oxidant and is a major component of photochemical smog.

P

parasite	An organism that gets its energy by feeding upon another organism called the host. In a parasitic relationship, the host suffers.
pH	A measure of the amount of acidity or alkalinity of soil or water.
phosphorous	Essential nutrient for all life forms, not readily available for uptake in soil.
photosynthesis	The process carried out by the green plant pigment chlorophyll, which is capable of combining water and carbon dioxide to produce glucose and oxygen using light energy.
pioneer plants	The first plants to appear in a sequence of succession in a particular environment.
pollination	The process in which pollen gets from the stamens to the pistils of flowers.
pollution	Harmful wastes introduced into an ecosystem.
population	A group of organisms of a single species that inhabit a certain region at a particular time.
prairie	A large area of level or rolling land covered with few trees and dominated by grasses.
precipitation	Water deposited on the earth in the form of rain, sleet, hail, or snow.
predator	An animal that hunts other animals and feeds upon them.
prediction	The process by which scientists use information to attempt to tell what will take place at a later time.

Glossary

preserve	A protected area, or sanctuary.
prey	Organisms that are killed by other organisms and used as food.
producer	Any green plant that makes its own food using chlorophyll and light energy.
pupa	The intermediate stage that follows the larval stage in many insect life cycles. An adult insect emerges from the pupa. Plural is pupae or pupas.

Q

quadrat	A convenient square or rectangle laid out for studying the relative abundance of populations.

R

rainforest	A closed, multi-storied, unbroken canopy of broad-leaved vegetation characterized by diverse life forms. Tropical rainforests feature an exceptional range of flora and fauna. More than half the earth's species are found in rainforests.
renewable resources	Natural materials or energy sources that may be renewed or replaced. Forests and solar energy are examples.
recycling	Using products again in new ways.
regional haze	Reduced visibility due to the formation of tiny particles of sulfur dioxide.
reserve	A preserve, a place in which all wildlife are protected from hunting or removal.

S

savanna	Ecosystem between a dry tropical forest and grassy plains.
satellite images	Pictures taken from satellites far above the earth's surface.
scavenger	An organism that feeds only on weak or dead organisms.
semiconductor	A thin sheet made of silicon and other elements that can change light energy into electricity.
sink	Reservoir.
smog	A brownish haze that occurs often in the summertime.

Glossary

solar	Having to do with the sun.
soot	Particles of dirt or dust which are often acidic in nature. These are emitted by the burning of fossil fuels.
stability	Steadiness; the ability of a community to return to its original condition after experiencing a change. Climax communities have a high degree of stability.
static	Showing little or no change.
stratosphere	The region of the atmosphere which is 30–40 kilometers above the Earth's surface.
succession	The orderly and predictable change in a community over a long period of time. The change is usually from a very simple community to a more complex one.
swamp	Wet, spongy land often populated with shrubs and trees.

T

taiga	The coniferous forest that stretches across northern North America, Europe, and Asia. This is a type of ecosystem.
technology	Applied science; engineering.
temperate	Climate characterized by warm summers and cold winters. Temperate climates are mostly found at mid-latitudes.
topography	The surface features of a region, which include hills, valleys, lakes, rivers, and so on. Topography is considered to be an abiotic factor.
topsoil	The uppermost layer of soil that contains minerals, water, and air spaces. Plants grow well in fertile topsoil.
toxic	Poisonous.
transpiration	The process of giving off water vapor through the leaves of green plants.
tundra	A type of level or treeless plain in the Arctic and Antarctic regions. The plants consist mainly of mosses, lichens, and grasses.

U

ultraviolet radiation	Solar rays that produce a sunburn and can cause cancer of the skin.
understory	Forest plants below the canopy.

Glossary

W

weathering Those processes that tend to break up or chemically dissolve solid rock.

wetlands Swamps, bogs, and marshes that are covered by plants adapted to living in wet environments.

INDEX

A

Abiotic factors, 4, 5, 17, 33
    influences on, 35
    studying effects of, 36
Acid deposition, 101
    dry, 103
    wet, 103
Acid precipitation, 101.
    See also Acid rain studying, 106
Acid rain, 103, 104–105, 106
    in 1999, *104*
Aerosol, black soot, 116–117
Agriculture, ecological effects of modifying, 163
Aquifers, 101, 141
Arsenic, 141–142
Atmosphere, 41
    acid particles in, 103
    industry and, 121
    nutrient atoms in, 69
    pollutants in, 101, 102.
    See also Pollution; specific pollutants
    rock exposure to, 82

B

Bacteria, 50, 62, 70, 74.
    See also Decomposers
Bare rock succession, *81*, 81–82, *83*
Biotic factors, 2, 4, 5, 17
    effect of shade on, 36, 37
Biotic system, 33
Body weight, optimum, 17
Bromeliads, 60

C

Canada geese
    Mississippi Flyway population of, 17–20, *20*
    on wildlife refuge, 18
Canopy, 86, *91, 93*
Carbon cycle, 69, *70,* 70, *114*
Carbon dioxide, 69, 113, 114, 122, 151
Carbon monoxide, 121
Catalysts, 124
Catalytic converter, 121
Cattle
    dependence on plants of, 3
    grazing, *1*
Celsius scale, 6
Chlorine, 135
    atoms, effect of, on ozone, *135*
Chlorofluorocarbons (CFCs), *135*
    in stratosphere, relative increase/decrease of, *137*
    threat of, 137
Chlorophyll, 49
Clean Water Act of 1972, 19
Climax community, 79–80
    oak-hickory, *80*
Commensalism, *60,* 60
Community
    Arizona desert, *27*
    balanced, 79
    bog, *29*
    change, 80
    defined, 49
    emergent pond, *28*
    Florida pine flatwoods, 28
    influence of abiotic factors on, 35
    milkweed flower head, *29*, 30
    natural biological, 30, 33
    niches in, 49
        forest, *51*
        prairie, 63
        Oklahoma, *31,* 49
    relationships, 58–63.
    See also specific relationships
Community concept, 30
Competition, 41, 63, 82
    interspecies, 63, 82
    intraspecies, 63, *64,* 82
Conservation, water, 145, 146
Consumers, 56
    in forests, 79
Cubic units, 9
Cultures
    changes within, 24
    setting up, 23

D

DDT, 19
Decomposers, 50, 62, *62,* 70.
    See also Decomposition
Decomposition, 49, 50, 113.
    See also Decomposers
    process, in Gulf of Mexico, 142
Deforestation, 40
    causes of, 41–43
Density
    changes in, 11
    defined, 9
    estimating, 12–13
    measuring population, *14*
Dependence, 3
Desalination, 141
Desert countryside, 33–35, *34*

Index

Direct observation, 5
    tips and suggestions for, 6
Drought
    in Florida, 2001, 141
    in United States, 1988, 145
Dynamic balance, 35

E

Eater-eaten relationships, 57
    constructing food web showing, 65
Ecologists, 1
    concept of community, to, 30. *See also* Community; Community concept
    predictions of, 81
    study
        areas of, 2–3
        of ecosystems by, 35
        of interactions by, 4
        of succession by, 86–87
Ecology
    basic business of, 35
    defined, 1, 4
    effect of agriculture on, 163
    organizing study of, 30
    study of natural relationships in, 3, 4
Ecosystem, 33, 35
    closed, 69, 75
        creating, 76–77
    destruction, 163
    effect of industries on, 121
    endangered, 39, 47
    importance of trees to, 44
Electricity, 124
    collecting methane gas for, *157*
Electrons, 124
Emigration, 11, 17
Emissions, gas, 102, 123
    smog-causing, 122, *122*
Endangered species, 19, 39. *See also* Extinction

    of trees, 41
    in Tropics, 44
Energy, 49
    fuel cell, 124
    in industrialized countries, 130
    light, 69, 116
        effect of, on aquarium water, 36, 37
    solar, 125, 130
    use, in the United States, 101
Energy pyramid, 56
    for pond food chain, *57*
Environment, 33
    abiotic, 69
    change of, by cities, 54
Erosion, 81, 82, 101
    of topsoil, 163
Exhaust, automotive. *See also* Fossil fuels; Motor vehicles; Pollution
    effect of, on plants, 36, 37
Extinction, 19. *See also* Endangered species
    and monarch butterflies, 46
    tree species close to, 41

F

Fahrenheit scale, 6
Fertilization, 43
Flaring stations, 151, *152*, 156
Florida Everglades, 47
Food, 49, 101
Food chain, 50
    common, *55*
    constructing, 53
        in a city, 54
    energy loss in, 56
    energy pyramid for pond, *57*
    investigating human, 55
    movement of chemicals through, 69
    as nutrient cycle, 69
    prairie, *52*

Food-energy transfer. *See* Food chain
Food web, 50
    constructing, 53
    nutrient atoms in, 69
    showing eater-eaten relationships, constructing, 65
Forests, 41, 42. *See also* Rain forest; Trees
    effect of bird migration on, 79
    and farms, American (1860-2000), *43*
    fires in, 80, 81, 117
    mangrove, 39
    northeastern U.S., 42–43
    oak-hickory, 79, *80*
    destruction of, 80
    promoting regrowth of, 44
    quadrats in, 86
    reasons to preserve, 44
    tropical
        destruction of, 40
        dry, 39–40
        rain, 39
Fossil fuels, 130. *See also* Motor vehicles
Fresh Kills landfill, 151, 156
Fruit flies
    collecting data on, 24
    life cycle of, 22, 24
    setting up cultures for, 23
    studying population of, 22–24
Fungi, 50, 62, 70, 74. *See also* Decomposers

G

Garbage dumps, 151, 159
Gases, 113–115, 122, 123
Gasoline, 121, 122. *See also* Fossil fuels
    consumer reduction of 123

Genetics, 22
    use of, to improve plants, 43
Germination, 46
Giant sequoia, 41, *42*
Global warming, 41, 101, 109–110, 121. *See also* Greenhouse effect
    causes of, 113–115
    effects of, 111–113
    graph, *109*
Grazing, 79
    buffalo, *31*
    cattle, *1*
    effect on savannas of, 39–40
Greenhouse effect, *116,* 116–117. *See also* Greenhouse gases
    demonstrating, 118–119
Greenhouse gases, 113–115. *See also* Greenhouse effect
Gulf of Mexico, dead zone in, 142

H

Habitat, 12
    optimal, 18
Herbicides, 163
Homeostasis, 17–18, 21, 35, 49, 56
Homeostatic balance, 82. *See also* Homeostasis
Hormones, 21
Host, 58
Humans
    carrying capacity, 99–100
    as change agents, 4, 19
    importance of role of, 163
    population growth over 18,000 years, *99*
    as threat to wetlands, 47
Hunting
    effect of, on quail population, *15*
    geese, impracticality of, 19

I

Immigration, 11, 17
Incineration, 152, *158*
Industrial Revolution, 99
Inorganic material, 69
Insecticides, 19. *See also* Pesticides
Interactions, 4, 33, 49, 82
    changes, as evidence of, 5
    in communities, 79
    community of organisms, involved in, 30
    in goose population, 17–18
    predicting future environmental, 13
    sample diary for, 6
Interspecies competition, 63, 82. *See also* Competition
Intraspecies competition, 63, *64,* 82. *See also* Competition
Irrigation, 43

L

Landfill gas, 151, 156
Landfills, 151
    capping, 156
    sanitary, 151–152
    Oceanside, New York, 156, *157, 158*
Larva, 22, 24
Leaching, 152, 156
Lichen, *61,* 61, 82
Life cycle, 69
Lipids, 69
Locomotion, 14

M

Marsh, egret standing in, *47*
Melt pool and iceberg, *112*
Metamorphosis, 22
Methane, 115, 151
    use of, to produce electricity, 157

Migration, 11, 18
    effect of bird, 79
Molds, 50, 62.
    See also Decomposers
Monarch butterfly, 44
    danger of extinction to, 46
    See also Extinction
Mortality, 11, 17
Mosses, 82
Motor vehicles, 121–125
    hybrid, 123
    solar cars, 125
    sport utility vehicles (SUVs), 121, 122
Mutualism, 61, *61*

N

Natality, 11, 17
Niches, 49
    consumer, 55
NIMBY (Not In My Back-Yard), 152
Nitrogen oxides, 102, 115, 121
Northern Lights, 117
Nutrient atoms, 69
Nutrient cycles, 69, 75
Nutrients, 69

O

Organic material, 69
Organisms, 1, 4. *See also* Biotic factors
    community of, 30
    dead, 50
    decay, 74. *See also* Decomposers; specific organisms
    in food webs, 69
    homeostatic balance of, 21
    interacting, 33
    population of, 9
Overpopulation, 99
Ozone, 101, 116–117
    defined, 133

Index

effect of chlorine atoms on, 135
gas, 102
loss, 135–136
relative amount of, over Antarctica during month of October, *137*
Ozone cycle, 134, *134*
Ozone layer, 101, 133
depth of, *136*
hole in, 136
illegal CFCs in, 138

P

Parasitism, 49, *58*, 58
Pesticides, 163. *See also* Insecticides
pH, 106
measuring, 6
Phosphorus cycle, 74, *74*
Photosynthesis, 41, 44, 113, 163
Photovoltaic (PV) cells, 125. *See also* Solar cars
Pioneer trees, 80. *See also* Forests
Polar bears on ice, 111
Pollutants, 105. *See also* Pollution
need to reduce marine, 163
Pollution, 25, 100, 102, 121. *See also* Pollutants
air, 101, 105, 122
controls, 121
calculating, in business district, *128–129*
of oceans, 163
Population
changes in, 11
density, 9, 11, *14*
effect of hunting on quail, *15*
fluctuation in, 17
interacting, 33
Prairie, 30, 86
becomes farmland, 43

community, 63
Oklahoma, 31, 49
food chain, 50, *52*
predator, *32*
recycling nutrients back to soil of, *32*
Precipitation, 4
Predators, 18, 19, *32*, 59
loss of natural, 163
Preserves, 44
Producers, 56
Protons, 124
Pupa, 22, 24

Q

Quadrat, 11, *86*
canopy species
abandoned 100 years, *88*
abandoned 20 years, *90*
abandoned 50 years, *92*
ecologists' use of, 86–87
seedlings present, abandoned 5 years, *89*
understory species
abandoned 20 years, *91*
abandoned 50 years, *93*
Quail, 1, 2, 21
population, effect of hunting on, *15*

R

Rain forest
as climax community, 79
tropical, *39*, 39, 44
Recycling, 151
Regional haze, 101, 102
Relationships
community, 58–63. *See also* specific relationships
investigating special, 66
Reserves, 44
Resources, limited, 101
Respiration, 44, 113

S

Satellite images, 40
Savanna, 39–40
Scavenging, 49, 59, *59*
Semiconductors, 125
Shantytowns, 100
Smog, 101, 102, 103, 121
Soil, 2
building, 82
erosion, 81
forest, after fire, 80
observing wild plants in, 36
studying organisms in, 37
Solar cars, 125. *See also* Motor vehicles
testing model, *126–127*
Solid wastes
investigating disposal of, 153–155
problem of, 151–152
Square units, 9
Stability, 17
Stratosphere, 102
CFCs in, 135, 136
defined, 133
ozone in, 134
relative increase/decrease of CFCs in, *137*
Succession
forest, 86–87
patterns, common, 82
pond, *84*, 84–85
studying, 97
Sulfur dioxide, 102

T

Taiga, 80
Technology, 99
Temperature, 4
air, 35
increased, 111
comparing, 36
life cycles, dependent on, 22
maintaining constant body, 21

measuring, 6
moderation, by trees, 44
rising, due to global
   warming, 41
Topsoil, 41, 101
   capping landfills with, 156
   erosion of, 163
Toxins, 101, 103
   in landfills, 152
Trees, 44. *See also* Forests
   effect of acid deposition
      on, 103
   observing, 45
Tundra, 80

U

Ultraviolet (UV) radiation,
   133, 134, 136
Understory, 86, *91, 93*

W

Water, 141–142
   conservation, 145, 146
   consumption
      determining, 145–146
      home, *145*
   contaminated, 141, 142
   purification, 142
   systems, and problems from
      old dumps, 151
Water cycle, 72, *72*
   do-it-yourself, 73
Water lettuce
   estimating density
      of, 12–13
   on Florida swamp, *10*
Weathering, 82
Wetlands, 47, *47,* 115
World Wildlife Fund, 40
Worms, 70, 74

Share Your Bright Ideas with Us!

We want to hear from you! Your valuable comments and suggestions will help us meet your current and future classroom needs.

Your name_____Date_____

School name_____Phone_____

School address_____

Grade level taught_____Subject area(s) taught_____Average class size_____

Where did you purchase this publication?_____

Was your salesperson knowledgeable about this product?   Yes_____   No_____

What monies were used to purchase this product?

   ___School supplemental budget   ___Federal/state funding   ___Personal

Please "grade" this Walch publication according to the following criteria:

   Quality of service you received when purchasing .... A   B   C   D   F
   Ease of use... A   B   C   D   F
   Quality of content... A   B   C   D   F
   Page layout .... A   B   C   D   F
   Organization of material .... A   B   C   D   F
   Suitability for grade level .... A   B   C   D   F
   Instructional value.. A   B   C   D   F

COMMENTS:_____

_____

What specific supplemental materials would help you meet your current—or future—instructional needs?

_____

Have you used other Walch publications? If so, which ones?_____

May we use your comments in upcoming communications?   ___Yes   ___No

Please **FAX** this completed form to **207-772-3105**, or mail it to:

   Product Development, J. Weston Walch, Publisher, P.O. Box 658, Portland, ME 04104-0658

We will send you a **FREE GIFT** as our way of thanking you for your feedback. **THANK YOU!**